# THE GREAT FLOOD OF 1607

BARB DRUMMOND

BARB DRUMMOND

Copyright © 2024 by Barb Drummond

All rights reserved.

No part of this book may be reproduced in any form or by any electronic or mechanical means, including information storage and retrieval systems, without written permission from the author, except for the use of brief quotations in a book review.

Published 2024

Barb Drummond has asserted her right to be identified as the author of this wok in accordance with the Copyright, Designs and Patents Act 1988.

All images copyright the author.

Cover design by Derek Bainton

978-1-912829-14-9 The Great Flood of 1607 Paperback

978-1-912829-15-6 Hardback

978-1-912829-16-3 ebook

# CONTENTS

| | |
|---|---|
| *Introduction* | vii |
| 1. Before the Flood | 1 |
| 2. Records on the Ground | 10 |
| 3. God's Warning To His People | 24 |
| 4. Lamentable News | 42 |
| 5. Seyer's Memoirs | 56 |
| 6. Baker's Account | 64 |
| 7. Other Sources | 84 |
| 8. The Great Flood Tracked to Cardiff | 90 |
| 9. The Flood Tracked to Gwent Levels | 104 |
| 10. Why Not Wye? | 116 |
| 11. Severn Tidelands | 122 |
| 12. Severn Crossings | 146 |
| 13. Gloucester Storms | 156 |
| 14. Bristol and Its Hinterland | 162 |
| 15. The Great Flood in Somerset | 168 |
| 16. Barnstaple and Beyond | 180 |
| 17. Conclusion | 190 |
| *Bibliography* | 197 |
| *Index* | 205 |
| *About the Author* | 219 |

'God's Warning to His People of England'

"Upon Tuesday, being the 20th of January, 1607, there happened such an overflowing of waters, such a violent swelling of the seas, and such forcible breaches made in the firm land in the counties following ... Gloucester, Somerset, Monmouth, Glamorgan, Carmarthen, and divers and sundry other places of South Wales, the like never in the memory of man hath never bin seen or heard of. For about nine of the clock in the morning, the same being fayrely ceive afar off as it were in the elements huge and mighty hilles of water tombling one over another in such sort as if the greatest mountaynes in the world had overwhelmed the low villages or marshy grounds. Sometimes it dazzled of many spectators that they imagined it had bin some fogge or miste coming with great swiftness towards them, and with such a smoke as if mountaynes were all on fire, and to the view of some it seemed as if mylions of thousands of arrows had bin shot for the all at one time. So violent and swift were the outrageous waves that in less than five hours space most part of those countreys (especially the places which laye low) were all overflown, and many hundreds of people, both men, and women, and children, were then quite devoured by those outrageous waters — nay, more, the farmers and husbandmen and shepherdes might behold their goodly flockes and sheep swimming upon the waters dead. The names of some of the towns and villages which suffered great harmes and losses hereby were those viz., Bristoll, and Aust, all the countreyes along both sides of the Severn, from Gloucester to Bristol, Chepstowe, Goldclift, Matherne, Caldicot Moores, Redrift, Newport, Cardiffe, Swansey, Laugharne, Llanstephen. The foundations of many churches and houses were in a manner decayed, and some carried quite away, as in Cardiffe, in the countie of Glamorgan, there was a great part of the church next to the waterside beaten down with the water. Divers other churches lie hidden in the waters, and some of them the tops are to be seen, and some others nothinge at al to be seen, and some the steeples, and some of them nothinge at al."

D.M.C., Harleian Miscellany

# INTRODUCTION

I enjoy visiting historic sites, and many years ago I discovered a church in South Wales with a marker recording the height of "the Great Flood" of 1606/7, about 5 feet from the ground. It drowned huge areas of coastal Wales and the West of England. In the small parish of Goldcliffe a bronze plaque commemorates the deaths of 22 people. If you multiply this number across the region, we are looking at a significant death toll, plus huge losses of livestock and crops, and substantial property damage.

From another source comes the following: "The Avon, upon which Welford stands, ... was nearly drowned by its flooding in July 1588, when the Armada was in the Channel, and as we said: "God blew with His winds and they were not," and Shakespeare was 24 years old. That July storm which scattered Philip's ships would seem to be referred to in 'A Midsummer Night's Dream', not in regard to its greater business, but as having flooded out Welford.

"The winds, piping to us in vain,
As in revenge, have suck'd up from the sea
Contagious fogs, which, falling on the land,
Have every pelting river made so proud,
That they have overborne their continents.

> The ox hath therefore stretch'd his yoke in vain,
> The ploughman lost his sweat; and the green corn
> Hath rotted, ere his youth attain'd a beard:
> The fold stands empty in the drowned field,
> The crows are fatted with the murrion flock;
> The Nine-Men's Morris is fill'd with mud;
> and the quaint mazes in the wanton green,
> For lack of tread, are undistinguishable..."""[1]

This is terrifying, and may have been used by the Bard purely for dramatic effect. But what if he was writing from life? Welford-on-Avon is in Warwickshire, to the west of Shakespeare's home of Stratford. It lies in a large loop of the river, so floods must have been a frequent risk in winter. But it is in the heart of England, far from the coast and the Armada, suggesting the 1606/7 flood may have had an impact far inland.

What if Shakespeare and his peers had first-hand experience of extreme weather in an age when this was common? Welford is also described as an island in a great sea of meadow and orchard. This could easily apply to much of the land that borders the often-flooded River Severn.

One of the most famous images of the Elizabethan Age is that of the ice fairs on the Thames, in 1600–1814 as part of the Mini Ice Age which struck the Northern Hemisphere. The year 1683/4 was the harshest winter in memory across Europe. Lake Constance froze. John Evelyn described it in his diary, from 27 December as the most severe frost ever, and an ice fair where hogs were roasted, a printing press was set up to provide souvenirs, and a settlement was built that included a brothel.

This sounds like people were having a lot of fun. But there must have been a lot of suffering. Frozen water is not drinkable. People — especially the poor, the very old and the young — died of what is now known as hypothermia. Ships could not sail, roads were blocked, trade ceased. Wood was too hard to chop, the earth too hard to cultivate and many animals died, so suffering was widespread.

It seems the Great flood was not an isolated disaster. It arrived in the midst of decades of extreme weather in Europe. But it was the most spectacular as it carried the weight of God's judgement on his people, which added to the horror, and possibly delayed the recovery of the region and of the victims.

# CHAPTER 1
# BEFORE THE FLOOD

The Glastonbury lake villages approximately coincided with the start of the Christian era, suggesting milder weather attracted missionaries to these islands, so it was apparently a pull rather than a push. The monks were skilled in agriculture and engineering which allowed them to drain the low-lying lands to expand settlements.

The Bristol Channel was also formed about this time. The River Severn and her tributaries carry large amounts of particles, mostly mud which fertilises farmlands. This was necessary as few trees grew there to provide firewood, so local people often burnt dried animal dung, frequently mixed with straw, for cooking and heating. About 80,000 tons of silt are carried downstream by the Severn and her tributaries each year, enough to build up the land level by 6–7 feet if spread evenly. Its regular flooding has helped to build up the waterside lands often known as 'wastes', i.e. regions which are sparsely settled but highly productive in agriculture, fisheries and animal husbandry. The term 'waste' has a long history, defining regions that were not improved and/or built upon, such as what became Bristol's Queen Square. It was also applied to lands 'discovered' by Europeans to justify their seizure of them.

Before ships sailed, crews said prayers for their safe voyage, and on their safe return gave thanks to God and donations to maintain churches. Hailes Abbey was founded in 1246 by Richard of Cornwall in gratitude for surviving a storm at sea.[1]

Up to a century ago, the head of the star-shaped tidewater region of Carmarthen Bay was about 1 mile wide, with well-timbered flats on either side, rich with grass and fast-flowing streams. "No matting weeds have time to gather or mud to settle in these translucent streams... and sweep in graceful curves through the fat pastures of the vale."[2]

Large areas of dunes existed between Freshwater Bay in Pembrokeshire and Merthyr Mawr in Glamorgan. They seem to date from the early Iron Age, c.400 BCE, up to the Roman period. At Pennard on the Gower, the first castle was built about 1100; by c.1270 it was replaced and St Mary's Church was built. It seems the dunes existed in 1316 and by 1478 were recorded as a threat to settlement. The church was buried by 1528, but located again in 1861. The old church of St Nicholaston was also engulfed. The east of Swansea suffered worse encroachments. By 1205, dunes were recorded, and extended far inland by 1344.

Margam Abbey was threatened from 1336, mostly by losses to the sea; by c.1300 its hermitage was buried in the sand. Port Talbot was also engulfed, as an upright stone has been unearthed dating from 1626. The Via Julia seems to have passed through the north of the dunes and through the Kenfig Burrows, and a river crossing probably survived into the 9th century. There seems to have been no danger in the 12th century, as the castle was still in use in 1403. But by 1485 the king's highway to the north east was being engulfed and the region was often beset by storms.[3]

Gerald of Wales wrote of a storm in winter 1171/2 in what is now Newgale, St Bride's Bay which "laid bare... the surface of the earth, which had been covered for many ages, and discovered the trunks of trees cut off, standing in the very sea itself, the strokes of the hatchet appearing as if made only yesterday... By a revolution, the road for ships [i.e. channel] became impassable, and looked, not like a shore, but like a grove cut down... being by degrees consumed, and swallowed up by the violence and encroachment of the sea".[4] Thomas Pennant made similar claims of trees on the foreshore of Abergele near

Denbighshire: "The wood is collected by the poorer people and carried home and used as fuel".⁵ This was in a region known for its shortage of woodlands, so the story makes sense. There was also evidence of a sunken forest when Cardiff Docks were being built, and at Port Talbot, peat beds were found 44 feet below the local flats. The same was found at Cornwall where deposits included tin and human skulls.⁶

In c.1625 Richard James wrote:
"In summer places when ye sea doth bate,
Down from ye shoare, 'tis wonder to relate
How many thowsands of thiese trees now stand
Black broken on their rootes, which once drie land
Did cover."⁷

To the far west in Pembrokeshire, desolate dunes, known as burrows, stretch inland, rising to 200 feet high towards the fertile valley of Castlemartin.⁸ But North claims the term sunken forests should be replaced by the term 'sunken lands', as some of the peat layers are collections of vegetable debris and marsh plants that may have drifted into shallow water where they sank.⁹

Historically, sea levels have varied widely in response to temperature changes, with extreme cold causing the formation of ice and snow on low-lying land, which lowered sea levels, especially at the poles. When the waters withdrew, the land dried out, so it rose again, and more land became available for agriculture and settlements, which were less subject to flooding. The historic height of this rise seems incredible. Shells have been found at the mouth of the Bristol Channel which would usually be found at a depth of nearly 60 fathoms.

Along the south east coast of Wales is a region with the name Wentloog, which suggests links with the Low Countries which in the 17th century provided these islands with engineers to drain the coastal levels. But this term derives from Llansanffaid Gwynllwg, the extensive peatlands which comprise much of the region. The landscape becomes linked with the Romans, based at Caerleon from where legions carried out much of the coastal drainage, but whose port is now lost beneath the waves. Over the ensuing centuries, walls and defences have been built, repaired and replaced.

In South Wales, the best-known floodplains are the Wentloog and

Caldicot Levels, between Portskewett and Rumney, the modern Severn Crossings and the east of Cardiff which mostly date from the end of the last Ice Age. But they are below the levels of high spring tides so would still be at risk of flooding if they were not protected by flood banks. At the present time, the amount of deposition is about equal to erosion, so the system seems stable, but with the increase in sea levels from global warming, this could become a major concern. North claims that sea levels at the end of the last Ice Age were relatively lower than now, so areas of marsh and woodland extended into much of what is now the Bristol Channel.

The Great Flood of 1606/7 is often described as a storm, or more recently a tsunami. It swept in from the west, so the following history will start in the far west of Wales, travel up the River Severn, then follow the English coast southwards.

Normans settled as far west as Pembrokeshire and a source claims the wife of Henry I(1068-1135) was from Flanders; this seems to be his second wife, Adeliza of Louvain. Henry settled Flemings who had been made homeless by floods in their homeland. They were given asylum, but at a price: they were expected to defend the borders against the Welsh, which explains why so much of the region still speaks English. Another source claimed they arrived in 2 or 3 batches, to become so numerous that some were settled on the coast of Carmarthen and the Gower where they survived as pockets of English speakers such as in the small village of Flemingston. Wikipedia claims that in 1317 Flemingston was held by Philip le Fleming, so after the medieval warm period c.800-1300 and the start of the Little Ice Age. Further evidence of the Flemings comes from a list of 12 knights in Glamorgan includes Sir John Flemynge of Wenvoe, Lamays and Flemingston.10 Robert Fitzhamon granted St Donat's to the Stradlings, a family that survived for several hundred years. The first of them was Le Esterling, which suggests he was another refugee from the Low Countries. Another refugee there was Nicolas Breakspeare who became the first English pope, Adrian IV. Yet another sighting of Flemings is at Wiston, where a castle was founded about 1100 by a Fleming with the wonderful name of Wizo.[11]

These 'Flemings' thus provide us with some surprisingly modern history: of people apparently driven from their homes by extreme weather, forced to flee to safety in small boats. Were they the first climate refugees to arrive on these shores, or the first to leave records?

At Dinas in Pembrokeshire, a church built at sea level at Cwm-Yr-Eglwys "was overthrown in the Great Storm" of October 1859 when the Royal Charter ship was lost off Anglesey.[12] The central churchyard is now protected by a sea wall, suggesting it was prone to storm damage, and like many other sites has lost any evidence it probably had of the 1606/7 flood.

Laugharne is between Carmarthen and the coast; it seems to be the most westerly site recorded as a victim of the Great Flood. Between the town and the coast is an area which had long been subject to considerable storms and shifting sands.

The traveller and royal clerk Gerard of Wales (c.1146–c.1223) wrote of a winter storm in Pembrokeshire which exposed ancient trees, claiming axe marks were fresh as if recently made and which provided plentiful firewood for the poor. But these same trees blocked the 'road for ships' so destroyed shipping in the port and bankrupted merchants.

Swansea Bay is now a huge expanse of golden sand lined by cottages, which rivals Bristol for having Britain's highest tidal swings. Plans were made to exploit this with a tidal barrage to generate green energy, but this failed to attract funding. Historically, it was so full of sand it became a forest which was drowned at the end of the last Ice Age. Stories are told of the bay being farmed up till the 17th century, so about the time of the flood, and extending to the "Green Grounds of Mumbles Head", suggesting the area was fertile.[13] As recently as 1797 the bay was still described as boggy ground "at the mercy of the gales and high water".

The sea continued to cause problems; as recently as 1818 people still spoke of the area between the flat land at the Mumbles to the low water mark being land "before the sea made its inroads on the flat which now forms the admired bay of Swansea". A map is mentioned which showed the "ancient lands under the sea". Surprisingly, this area is recorded on the Cassini Ordnance Survey map of 1830/31, which shows what appears to be sands with streams taking up about half the modern bay, the remainder being labelled as Inner and Outer

Green Grounds which stretched past Mumbles Head into the main channel.

Thus the Great Flood was one of many extreme weather events; an account of the 1606/7 flood claims that at Swansea "many great harmes were done", but unfortunately fails to provide details.[14] It seems the 1606/7 flood thus caused permanent loss of ground, forcing locals to relocate inland.

Other areas on the south west coast of Wales suffered from sand being driven inland by storms, especially on the Gower Peninsula from 1554. An Act of Parliament was passed that year to deal with "the hurt, nuisance and losses by reason of sand arising out of the sea, and driven to land by storms and winds, whereby much good ground is covered, especially in the county of Glamorgan".[15] Two churches were abandoned, at Penmaen on Oxwich Bay on the Gower and further inland to the east at Pennard. By 1650 land close to the sea owned by the lord of the manor was ruined by sand encroachment, so was abandoned to become common ground.[16] This shows the region struggled against storm damage for many centuries and that the 1606/7 flood was in the midst of a period of extreme weather. But this long struggle probably caused depopulation of the area, leaving fewer people to be victims of the 1606/7 flood or for their records to have survived.

This is from Rhys: "[Kenfig] is a wild stretch of coast that runs westward now (1911) from the estuary of Ogmore River." Kenfig is near the coast west of Porthcawl, and it is hard to imagine it was a major international port with moorings for 24 ships, albeit they were small ones. It had a large pool which was said to be prone to flooding. But encroachment of sand forced the locals to build a new church inland, and by 1316 their pastures had been overwhelmed by driven sand.[17] A new road was built further inland and Leland described the settlement as being "devoured" by sands. This coincides with the surge in storms in the North Sea, so the problem was widespread. The extreme weather began in 1219 when the first St Marcellus Flood drowned many people in the northern Netherlands. It peaked in 1362 with the Grote Mandrenke, 'Great Drowning of Men' which carved out the Zeider Zee in what is now the Netherlands, and when unknown numbers of people died. It also created the harbour of Dunwich, in

Suffolk then capital of the East Angles during northern Europe's 'Mini Ice Age', which coincided with — or drove — the emigration of the 'Flemings'. Kenfig is now a small village still being eaten by the sea.

"The battle between sea and land was fought there with endless change of fortune: the sea hurls up billows of land to choke the fields and bury the houses; the land sends out deadly ridges of low rock to the murder of ships that pass in the Severn Sea.

"Newton Nottage stands back from the sea, with a broad belt of half-clothed dunes in between. The village is a survival of other days... The church, again, is original in its antiquity; and its tower is not like other towers, but has an 'air' not easy to realise as it stares across at the Merthyr Mawr warren like some amazed and amazing creature spawned by a primitive world".[18]

To the north east is the Kenfig Pool and Dunes National Nature Reserve, and east of Porthcawl is Merthyr-Mawr Nature Reserves, both barren sandy areas which may have been created at the above time. The Welsh coastal path now crosses the mouth of the former harbour. To the east is Cardiff airport, another sign of poor quality land.

Further along the coast was a Roman port which in the 2$^{nd}$ and 3$^{rd}$ centuries became the small settlements of West and East Aberthaw in the Vale of Glamorgan, on a safe natural harbour, though the latter has declined so is absent from some maps. Rhys describes it as a "legendary harbour" of South Wales which played a significant role in the Normans' arrival c.1093. Its wooded surrounds provided a safe mustering place for the invasion. One sources claims St Curig founded the church to pray for the souls of sailors, suggesting many ships were lost there.[19] The region is now scattered with small settlements, suggesting a general depopulation over time, as does the presence of a power station and cement works. Wikipedia claims it was a "safe, natural harbour where settlements existed from the 2nd century, and Roman pottery, coins, jewellery and tiles reflect its role as a port during the Roman invasion". A road led west towards marshland and a ford across the estuary. "By the 16$^{th}$ century the port of Aberthaw, south east of the village, was a thriving port, exporting wool and returning with wine, salt, dried fruit from Northern France", so it had a similar history to that of Bristol and other regional ports. It expanded when nearby Porthkerry harbour was destroyed by a storm, to become the main stopping point for ships between Cardiff and Swansea. By the early 17$^{th}$ century ship sizes had grown with the rising long distance trade to the West Indies, so the port must have expanded to accommodate them.

But these were long-term events, which allowed locals to adapt, to build embankments and drainage ditches, or to move to higher ground. As an author commented on the land beside the River Severn in Gloucestershire, these regions tended to be inhabited by the poor, who adapted to the extremes of changing weather whilst the great landlords lived safely inland on higher ground. When the 1606/7 flood struck, landlords were active in rescuing and relieving the victims, partly out of Christian charity, but also to ensure their own rentals.

In the nearby parish church of St Athan are 2 ornate, garishly painted tombs, claimed to be "the supposed burial place" of Sir Roger Berkoles in full armour at prayer who died in 1331 and his wife Katherine Turberville of Coity Castle. In the same side chapel is the tomb and "presumed burial place" of Sir William Berkoles(1302-27), son of Sir Roger of East Orchard and his wife Phelice De Vere. They reflect the huge wealth of the region, and of the collapse of this in the following centuries. The fact that such burials are merely "supposed" tells us much about the parish's decline.

There seems to be no records of the Great Flood causing major damage here, which fits with the general dearth of records. Possibly the tidal surge swept past the area, scouring away the coastal sands. The coastal path passes over the mouth of the now silted-up bay. Maps show moorland further inland.

Following the 'defeat' of the Spanish Armada, Elizabeth neglected the navy, and the Stuarts followed her lead. This led to a fall in trade, and so in customs income whilst merchants complained of the rise in raids on ports and shipping by England's many enemies. In a single year, Bristol merchants claimed to have lost £8,000 which forced them to invest in arming their ships. James I failed to act until 1620 when he began collecting Ships' Money. Trade with southern Ireland and South Wales was often attacked. Complaints were made of Algerian, Dutch and Turks "devouring English trade".

# CHAPTER 2
# RECORDS ON THE GROUND

The Great Flood struck a huge area, so it involved a wide range of geography, most of which was rural. Some people lived in hamlets too small to be named, so left no records. Tidenham parish runs between the rivers Wye and Severn down into the Beachley Peninsula over which the First Severn Crossing now soars. It has suffered extensive erosion from the Severn's high tidal range over the centuries, as reflected by the complete loss of 9 out of 24 churches since 625 AD, with 2 more surviving as ruins.

Accounts of the flood should make use of parish and civic records, but the quality and quantity vary widely. Margaret Walker wrote an article on "the church in sixteenth Century Swansea" which provides valuable details on the problems in finding records of the disaster. At the end of the 16th century, Swansea was yet another town under the rule of the wealthy Somerset family but the church of St Mary was the responsibility of the town authorities, unpaid officials who rarely had time for this often demanding role. A third of the costs to maintain the church were paid by the church, the rest by the town authorities, and many of their records survive for 1558–1761. This covers the time of the disaster, but the parish registers do not start until 1631, so no relevant burial records survive. Churchwardens' accounts, which record costs of the upkeep of the church cover 1558-1627, so should include reports of damage to the church and lands but there are gaps in their records for the relevant dates.[1]

The church was a Norman structure, so by then it was in poor condition, and there have since been 3 replacements. So if any marks were made of the flood level, they are long lost. Records survive of the purchase of 3 Welsh books in 1602, and a copy of "the Welsh omyly" published in 1606 was bought,[2] so close to the date of the flood. The only burials allowed within the church or chapel were of "those of the best sort" of the parish, so would not record most of the expected fatalities. In 1604 this included the vicar of Langewelache who had a knell rung for his funeral.[3]

Another issue related to the Great Flood is the accuracy or vagueness of the number of victims. Petra van Dam writes of the different ways to assess the physical impact of floods was to count the number of casualties and extent of damage, especially in financial costs, but reliable figures were rare across Europe. An area of Rotterdam was flooded, but the number of villages drowned exceeded those that existed. This is due to copies of accounts were often altered to include significant numbers from The Bible, so 33 villages represented the life span of Jesus.[4] Errors might also be introduced by published accounts having reached the printer via Chinese Whispers, illegible scripts, and by victims appealing for extra help for their plight.

The region that is best known as a victim of the 1606/7 flood is the coastal strip, or levels, from Cardiff eastwards to the modern River Severn road bridges which soar over the Newport/Caldicot Levels. The land is mostly low-lying, with some below the height of spring tides, so its existence still depends on the maintenance of the sea walls. It is still sparsely populated along the coast, though commuter villages and industrial parks are spreading. To the east of Cardiff, huge recycling plants loom above the speeding traffic; the rest is mostly a scrubby edgeland of mud, bungalows and travellers camps. Several churches still have markers which record the height of the 1606/7 flood levels.

The region's strange name of Wentloog suggests links with the Low Countries. The people of the region now forming the Netherlands and Belgium developed their engineering skills to drain and maintain their homelands and some came to these islands as early climate refugees. But the Victoria County History claims there was also a manor of this name at Kaeynsham, so another linked with English settlements which survived till the Dissolution of the Monasteries from 1536. This manor included the Monmouth parishes of St Mellons, Llanwern, most of the low country between Rhymney and the Usk plus an area of the hills. Part of what is now Cardiff was the manor of Roath Keynsham.

Wentloog also referred to an extensive forest which survived into the early 20th century when 2,244 acres were sold by Henry Somerset, 9th Duke of Beaufort. It was described as the largest wood in England, to the north east and partly within the boundaries of the city of Newport. It was also called Wentwood. There was a tannery at Tintern before 1241 which they bought bark for tanning leather from Wentloog for twopence per load.[5]

Normans colonised the Gwent Levels in the 11th and 12th centuries, attracted by milder climate and falling sea levels, and they built new sea walls before 1126. They founded 2 churches: St Brides/Bridget's in 1230 and St Peter's in 1142, in the parish of St Peters on the Moor, helpfully describing the land at the time. They drained the coastal area by digging the reens (a south western English term for a drainage ditch or channel) and ditches that still punctuate the landscape. But failure to maintain them, mostly due to Henry VIII's Dissolution of the Monasteries, caused the land to decline into tidal salt marshes.

Britain's population saw a massive fall during the Black Death but the churches received large donations to say prayers for victims' souls, so by the end of the epidemic this funded a huge Europe-wide founding and rebuilding programme of churches for the survivors. In the 1540s Leland wrote that the coast between Peterstone and St Brides became a medieval port and haven with wharves along a sea wall. Many of the gouts/drains of this time survive. But this seems to contradict the current Ordnance Survey map which shows a large rectangle beyond the sea wall. This land reclaimed from the sea beyond the present walls is called 'Warth'. It is washed by the tides, encouraging samphire and other salt-loving plants to provide grazing for lamb which became popular under the Stuarts. The Roman port is now further out, beneath the channel.

The Victoria County History claims "both churches [i.e. St Brides and Peter's] were restored again post 1607 flood and again by Victorians, so both are mixtures of styles". Both have markers showing the height of the flood, though St Brides is rarely open and St Peters is in private ownership so inaccessible.

The following account begins in the east, tracing records that survive in the region. My first awareness of the flood was when I visited Redwick church. Near the gate is a marker on the buttress about 4 feet high and another outside its porch about a foot higher which show the height reached by the floodwater. But following the Victorian rebuild, the latter may not to be its true height. Inside the porch is a copy of the dramatic print produced soon after the event, of church towers looming above the floodwaters while people and animals drowned.

The church at Redwick has a rich history and had several names, but has been returned to its original dedication of St Thomas the Apostle, the doubting Thomas of the New Testament. After the death of Christ, he allegedly went to India and is often associated with stormy weather, so this again raises questions about settlement in the region being driven by storms and the arrival of the Flemings further to the west.

At Whitson is a small redundant church, with no record of any dedication to a saint; it served as a family chapel for the landowners and their tenants. It is now in private hands and used for farm storage. The porch has been scraped of any details or plasterwork, a common practice of the Victorians, so though it is close to the coast, any records that may have existed have been lost.

Beyond Clifton Common are the Seawall tearooms, protected by the sea wall which is popular with Sunday strollers. On the wall is a farmhouse which includes the remains of Goldcliffe Priory which supervised the region's drainage. The stone used in the wall includes silica which glitters in the sun, hence the location's name. This is a fine place to appreciate the Severn and its sea defences, and the English coast beyond, a reminder of how close they were when most transport was by water. Turning inland, it helps to appreciate the flatness of the area and the drainage channels and ditches, some of which are marked with lines of poplars giving the region a Mediterranean appearance.

Inland is the Church of St Mary Magdalene at Goldcliff. It has never been open when I have visited, and there is no information board to provide contact details. But a book on Welsh monumental brasses provides details of a plaque measuring 20 x 8cm about 3 feet above the floor on the north wall of the chancel which records:

"1606 ON THE XX DAY OF IANUARY EVEN AS IT CAME TO
PAS IT PLEASED GOD THE FLVD DID FLOW TO THE
EDGE OF THIS SAME BRAS AND IN THIS PARISH
THEARE WAS LOST 5000 AND ODD POWNDS BESIDES
XXII PEOPLE WAS IN THIS PARRISH DROWND"

The plaque was erected by John Wilkins and William Tap in 1609, suggesting how long it took this small parish to recover to fund such a memorial. It also suggests the water rose higher at St Brides, but it is impossible to compare without knowing the height of the churches above sea level. This church is on a rocky outcrop, as opposed to St Brides where the flat churchyard is waterlogged and a corner of its tower is subsiding.

Inland from the Newport Wetlands is Nash church where a buttress bears a modern plaque recording the flood level, but with the modern-style date of 1607, which may be confusing, suggesting there were 2 floods.

To the west of this, beyond Newport on the River Usk, is the ancient Church of St Bridget at St Bride's, Wentloog. It is dedicated to an Irish virgin who, like St Thomas, was an early evangelist. Another church is dedicated to her to the north on the Monnow at Skenfrith which dates from the 12$^{th}$ century. Yet another is near Bridgend, all of which suggest they were founded by pioneering Christians, probably before the Normans built their grander edifices.

I had passed this church several times but never found it open, but I was able to gain access and was given a tour by a former churchwarden. She showed me the plaque showing the high-tide mark of the flood which fortunately had not been covered by subsequent layers of plaster, showing the importance to local people of preserving the record. It states "THE GREAT FLOOD 20 IANUARI IN THE MORNING 1606".

Some sources claim that the storm of 1703 was bigger. She told me the water level from the latter was a foot lower, but its marker has not survived. This later event, also called the Great Storm, is now recognised as a destructive extratropical hurricane which tore across eastern England, leaving vast destruction in its wake. On Goodwin Sands, where ships gathered to sail in safe convoys from the port of London, 1,000 men were lost. Defoe's book 'The Storm' claimed it was God's vengeance for England's lack of success in the War of the Spanish Succession. As with the 1606/7 flood, people sought reasons for the disaster, and blamed it on God's intervention. The 1703 storm struck London where 2,000 chimney pots were blown down, and the New Forest was ravaged, so it was the worst known storm to hit these islands, with estimates of 8,000 dead. Its more recent date, its scale and the damage caused means it is much better documented than the 1606/7 event, but this is the only record of it here, and it is significant that it reached the western region. John Evelyn lost many trees on his estate and was concerned at the harm this would cause to English shipbuilding, and the country's naval power. It was also different in that it was recorded as a storm rather than a flood, so lacked the Biblical connotations of victims being punished by God for their wickedness.

On 22 October 1707 a fleet was returning from the Battle of Malaga when a huge storm struck the fleet off Scilly Isles with the loss of almost 2,000 sailors, which was largely blamed on the fleet's inability to navigate accurately. This disaster inspired the Longitude Prize, the opening shot of the race towards improved navigation and many fields of science in these islands, and ultimately the Industrial Revolution. There is no mention of this storm causing damage in the region, but it must have reached there, though apparently without overwhelming the sea defences.

The former churchwarden at St Brides told me that the rheens/rhines had been poorly maintained, and a reservoir between the church and the coast had not been cleared, which must have played a role in the 1606/7 flood. This fits with the decline in knowledge and skills following the closure of the monasteries. We also discussed the risks to the region if another such flood were to strike, which seems increasingly likely as climate change becomes a greater threat to low-lying coastal regions. Church properties are especially at risk, as they cost so much to maintain whilst numbers of parishioners decline, and their average age rises.

Another plaque is said to be in the large church of St Peters at Peterstone, Wentlooge to the west of St Brides, on the coast. The church was built inland, but the sea has encroached since 1606/7. Unfortunately the church is now in private hands and signs warn of dogs on the loose to deter unwelcome visitors.

On the western edge of Cardiff is Rumney Great Wharf, occupied during the Bronze Age, and possibly also used by the Romans. Ordinance Survey maps show a big loop in the river which widens as it reaches the sea through gravel sands, a fish pool and allotments. The coastal path circuits it, and further east is a ragged coastal edge. It has been suggested that Rhumney Great Wharf was a significant settlement reliant on the sea in the past. The river formed the boundary between England and Wales.

Priests lived on the lands and collected rents, and were able to mobilise manpower to build and maintain flood banks, gates and drainage after the departure of the Romans. Controlled floods from the uplands deposited fertile soil which meant the land had less need of beasts for manure to fertilise crops. But when Henry VIII closed the monasteries, such large-scale land management ceased. Instead of migrating beasts between fields to prevent overgrazing, farmers often struggled to maintain their livelihoods. As the celibate clerical populations were replaced by small farmers with families, the population surged as yields declined. This surge in the population also drove the founding of 'plantations' in Ireland, to provide land for younger sons who in previous times would have entered the church. The same young men who fought in Ireland were at the forefront of Atlantic colonialism.

Another problem in researching the Great Flood is the confusion over dates. The Old-Style Julian system began its year on 25 March which fitted with the seasons and the annual agricultural hiring fairs.

It is continued by Britain's legal and tax systems. Britain changed to the Gregorian system in 1752, so at the time of the flood, it was in 1606 Old, and 1607 New Style, hence my writing the year as 1606/7.

The Great Flood is generally claimed to have swept up the English/Severn Channel, but records are uneven. Decline in church-going means many buildings have been lost or sold for private use, so any other flood markers may have been lost. The coast itself has shifted, especially along the ever-changing River Severn and the south west coast of Wales, where shifting sands have created some inlets and lost others and the coast is mostly in retreat.

Searching for an event at a known time should be straightforward, much easier than for family historians dealing with variations in spellings and overlapping generations. The various volumes of the epic Victoria County History allow online searches for words, but 'storm' scarcely appears, and 'flood' is virtually unknown outside of Somerset. Some of this can be due to records being lost over time. Parish records were often written on scraps of paper or vellum before being entered into registers, so could easily be lost or damaged. The books were kept in chests which may have been flooded, rendering the records illegible. Early collections of manuscripts reflected their owners interests, which could be narrow in focus, so they were inconsistent and their usefulness for modern researchers can be limited. The decline in learning Latin may also be a factor, placing potentially useful details out of reach for many researchers including myself. Parish registers can provide valuable information, but their survival and the inclusion of these records cannot be relied upon.

Westbury-on-Trym is to the north, but now a suburb of Bristol. Its register from 1579 records a baptistry at Shirehampton and its records state that the plague had not reached there. This shows they included details beyond the obligatory births, deaths and marriages. But this is followed by silence, with no further records till 1644.[6]

Bristol's prolific historian Reverend Samuel Seyer(1757–1831) claimed that in Bristol the Great Flood waters were 8 feet deep, reaching the entrance to the 'crowd', or crypt, of St Nicholas' Church, close to the river's edge in the city centre. But he provides no source for this. Records for the city should survive, but the 4-volume set of 'West

Country Churches' makes no mention of the disaster. It was published in the early 20th century, so well before the The Blitz destroyed many city and church records. St Peter's was one of the city's oldest churches, well above the floodwaters and was home to some of the wealthiest citizens, so their registers were extensive and should have mentioned the event. As civic leaders, many of them would have been involved in rescue and recovery after the Great Flood, so the apparent silence is disappointing. St Mary le Port was also above the high-tide mark. The parish of Philip & Jacob's dates from at least 1174, and is later referred to as St Jacob in the Market, suggesting its focus was towards Gloucestershire and the castle rather than the city. But like St Thomas and Temple, the parish was crowded with industries. It was often flooded in the 19th century, but by the Frome rather than the main River Avon. The parish of Temple must have been one of the worst affected. WJ Robinson claims the parish's records from 1558 were by 1916 in "an excellent state of Preservation" and covered a description of the Weavers' Chapel and water supply. In 1366 the plague struck, followed by several more times: in 1565 when there were 191 victims, in 1604, 168 and in 1645, 339 so there should be mention of the flood in this source.[7] Were floods — even that of 1606/7 — too common to be recorded in the low-lying parish? Or were the records kept elsewhere, organised centrally by the corporation who co-ordinated the city's rescue response?

St John the Baptist is the last surviving church built into the city wall. It is at the bottom of Small Street, close to the River Frome, so it must have suffered from the floods, especially of its crypt. But 'West Country Churches' claims "there seems to have been a general reconstruction of the chancel, following the original style, in the 17th century".[8] This suggests that it was flooded in 1606/7 but any markers of the flood have been lost and, it seems, so were the parish records of the time. But the following survives: "the remains of Sir George Snigge, the son of Baron S, recorder of Bristol are interred there" [i.e. in the crowd]. He was drowned in Dec 27 1610 while attempting to cross Rownham Ferry as he was returning from a visit to Sir Hugh Smythe at Ashton Court".[9] Their churchwarden accounts are in a good state from 1538 so perhaps the flood records here were ignored or omitted by editors.

The nearby St James Fair was held in its parish graveyard, now a park, on 25 July until the reign of Elizabeth, when it moved to September. The parish has some of the best records in the county from 1559, but there are no flood records.

Perhaps Ubley in Gloucestershire provides an answer with "the ... register contains on fly-leaves accounts of severe storms and other calamities". In 1683 "There was a mighty great frost, people did dye so fast".[10] Thus, flood records were probably scribbled on separate sheets, if written if at all, in the heat of the emergency and were subsequently lost. Britain — especially the western regions — is still famous for its rain, so were episodes of extreme weather seen as acts of god, useful only in sermons to warn against wickedness? Or were records not valued at all until the 18th century Enlightenment when gentlemen interested in natural history had the leisure to investigate them?

Britain also suffers huge regional variations in weather. Some regions, especially if low-lying ones, were so prone to floods that they were avoided by travellers, and the gentry lived on higher ground, so few records were made by first-hand witnesses.

Fred Hando quotes the 'Book Of Llandaff' when Gwaednerth murdered his brother Merchion, and was punished by 3 years' excommunication. But a new bishop pardoned him, and he "gave to God and Llandaff all the land, the woods and the sea coast of Llan Cadwaladr". This refers to the Wentloog Levels. But this seems not to have been much of a gift, as the "Llan reached as far as the Roman sea wall, and included all the 'Rotten Lands'".[11] Were these "rotten" lands formerly productive, but then neglected, the drainage skills lost at the Reformation? The presence of so many churches in close proximity on the levels shows the region was flourishing under the Normans. In particular, St Peter, the 'Cathedral of the Levels', is a large building of high standard. Its close proximity to the wharf suggests it was not at risk of flooding when built, as the Roman port is now far into the channel. Were these land grants a means of disposing of low value land and calling it an act of piety? This may reflect the practice of clerics — especially the Cistercian founders of Bristol's St Augustine's Abbey which had such a strong presence in South Wales.

This in turn poses more questions about the large numbers of churches that were built on low-lying riverside lands, most famously at nearby Tintern. These were often the only open space available in narrow valleys, but the location put the churches at risk of flooding whilst parishioners often lived on the safer slopes.

We also need to look at how much human activity has changed the landscapes since the Great Flood, often driven by the post-Reformation rise of populations in need of land to support themselves. Hando claims half of Sudbrook's embankments have been devoured by the sea in recent centuries. A camp existed there long before the Romans, apparently projected into the Severn near the first crossing to guard the nearby port of Chepstow. Surprisingly drastic changes occurred as recently as 1860, when 70-ton barges travelled ½ a mile upstream from the present viaduct to deliver coal via the tidal pill to the tiny settlement in Monmouth of St Pierre, a further mile and a quarter inland.[12] When I visited Portskewett, I stood in the waterlogged churchyard and was shown a nearby field with a small stream which a parishioner claimed had been a Roman port. In dealing with this flood, we have to adjust to the small populations at the time and of the wealth of goods being carried in what were, by modern standards, tiny boats. Also, most of these inlets were tidal, so a visit at low tide can be misleading.

Centuries of human activity in the region converge at the 2,000 year old Sudbrook Camp which is close to the 700-year-old, heavily eroded Trinity Chapel and the Severn Tunnel pumping station. The cost of digging Brunel's railway tunnel unexpectedly surged when workers hit a spring which caused flooding. They recouped some of their costs by selling the spring water to locals. The tunnel still needs high maintenance pumping to keep the trains running.

Human activity further declined in the area as by 1758 locals were no longer mowing grass at Sudbrook. But the churchyard was still in use for burials as Captain Blethin Smith of Sudbrook was carried to his grave by 6 seafarers in the late 18th century. But the area was not completely abandoned as the stone was valued for its ruggedness and was used in the construction of the lower part of Newport bridge.[13]

Before the modern state, there were no organisations dedicated to organising and providing emergency relief. This was largely left to the major landlords, so apart from a few parish records, it is mostly their estate records which preserve accounts of the 1606/7 flood. But these records were not written for the public or for modern research, but for their own accounting and legal purposes. Extreme events such as the flood are rarely stored in a way to make them accessible to modern researchers, and Landowner's archives are often on their country estates. This makes them difficult to access and to navigate, poorly indexed and with limited opening hours. Gloucestershire is uniquely fortunate to hold accounts of the Great Flood and its aftermath which can be found in the records of the sewer authorities.

An important, though limited resource comes from a memorial over the main inner door to the parish church of Hill, St Michael's which rises above the River Severn in south Gloucestershire. It was yet another church owned by St Augustine's of Bristol, from 1148 to the Dissolution, by the Jenner-Fust family from 1564 and the Fusts from 1609, so they had time to invest in it. They owned the property at the time of the Great Flood, but the parish registers date from 1629 so provide no record of the disaster or their response to it.

Bigland describes the now scenic village of Hill as "evil in winter, grievous in summer and never good for habitation",[14] which suggests Sir Francis Fust made spectacular improvements to his estate. He planned, built and erected the Great Sewer at Hill Pill, followed by repairing the church 9 years later. Mee claims "Both these public-spirited works have been of great and lasting value"[15]. Robinson adds this sewer was called the Imperial Draught, and that he built 2 others above it.[16]

Above the main doors of parish churches, there are often exhortations to behave well, to be good Christians, but at Hill, the words are more specific. Instead of listing the lord of the manor's noble ancestry or his gifts to local charities, this tells the story of his land improvements. There is no way of knowing whether this was to increase profits from the land, or to help his tenants. Most likely both. Memorials in the church suggest the family had a long history of involvement in worthy causes.

"To Drain this Parish from this During Flood / To Model and Repair this House of God / Are Patterns good set to Future Time / Free From yours ye Costs & Labour Mine.

Sir Francis Fust Baronet / Lord of this Manor for ye Benefit of its Inhabitants / at his own Expence Plan'd / Built and Erected in the Year 1759 the Great Sewer / at Hill Pill Next ye River / Call'd the Imperial Drought / and yeTwo Others above it. / He also in 1759 New Modled / And Repair'd this Church. / All the Costs and Materials / for all the Said Works / are as a Gift from him / Freely to this Parish for / the Use Above.

Let Those of Ability Strive to do Good"

The aged memorial is of what seems to be a simple, even primitive, design. The central region of text is bordered by what seem to be stocky columns, but on close inspection, they are tied with ropes, wider at top and bottom, representing pipes used in the drainage project. The monument is sited above the main inner door where some churches have the aforementioned exhortations or warnings to behave well. This also marks the memorial as unusual, as it aims to inspire worshippers to contribute what they can to the community. The apparent links with the Jenner family of vaccination fame are also interesting. Wealthy Georgians are often seen as lazy, corrupt, etc. But this family, as recorded by the memorial, seem to have spent their wealth on making the planet a better place whilst improving their own estates.

Bigland also describes attempts by a local landlord in recent times to try to prevent flood damage by the introduction of cord-grass, on the riverside of the sea walls, to convert the acres of tidal mud to meadows for grazing. It was introduced by a former squire of Hill, Jenner Fust, but its growth was leading to the decline of salmon which are claimed to dislike swimming in shallow waters.[17] The problem was worsened by the increased shipping to Sharpness which caused river pollution. A document in the Bodleian Library dated 1703, the only storm mentioned in Latimer's Annals of Bristol in the 18th Century, has a title which almost makes reading the rest redundant: "Fearful Newes of Thunder and Lightning, with terrible effects thereof, which Almighty God sent on a place called Olveston, in the County of

Glocester the 26th of November last." This was the Great Storm which Daniel Defoe wrote about that launched his literary career and is described elsewhere.

Hewlett describes extensive repairs in the aftermath of a bad storm on 4 November 1636 which caused serious damage to the sea wall at Hill, showing maintenance of the defences was an ongoing problem. Repairs were ordered to be made, "raising walls to their original height", suggesting the sea had been higher than the walls, so repairs had been inadequate.[18] Apparently the Civil War put an end to any repairs, so there was probably serious flooding to the region for many years.

## CHAPTER 3
## GOD'S WARNING TO HIS PEOPLE

The Great Flood was an unprecedented, terrifying event which struck a large area, and the flooding had profound Biblical significance, so it is worth looking at contemporary accounts. It seems the first off the presses was the pamphlet "God's Warning to His People of England By The Great Overflowing of the Waters Or Floudes lately hapned in South Wales, and many other places. Wherein is declared the great losses, and wonderful damages, that hapned thereby: by the drowning of many Townes and Villages, to the utter undoing of many thousandes of people". It was written by the Puritan William Jones of Usk, suggesting he was close to the site of the disaster, so it is reasonable to accept that his information was accurate. This language seems archaic, but it was echoed by Daniel Defoe a century later when the great storm of 1703 when "God chose the Storm to be a vehicle of his wrath, the means by which to chastise London... 'a powerful, popular, wealthy and most reprobate city'... to direct its inhabitants towards godliness and repentance".[1]

The pamphlet was entered into the register of the Stationers' Company of London, i.e. published and copyrighted, on 23 February 1606, a mere month after the event so it is the closest at the time to being 'hot off the press'. Yet as with many other sources, there is confusion over the year. The document is dated 1607, suggesting it was written a year after the disaster, which is clearly nonsense, so the calendar system needs to be clarified.

The Old-Style or Julian Calendar was established by Julius Caesar and it followed the agricultural seasons, with the year starting on 25 March. This was often the date for hiring fairs, as most agriculture began in spring. The New, Modern-Style or Gregorian Calendar came into effect in 1582 as a modification of the Julian, to make the length of the year more accurate, based on improvements in astronomy. It begins on 1 January but was seen by many Protestant countries as a Catholic system, so was not accepted in Britain and its colonies until 1752, and involved the loss of 10 days to allow for planetary movements. Hence the flood in question struck on 20 January 1606 Old Style (OS) but on 30 January 1607 New Style.

But it seems there were also local variations, as the records from St Mary's in Swansea show January or February were cited as the start of the year in the earliest accounts, but after a gap of 10 years, the records recommenced with the year starting in May, June or July.[2]

A further explanation comes from an article on the flood from the National Library of Wales which makes the calendar system almost beyond belief and throws into question many dates from the period with "Many readers and modern writers are unaware that previous to the year 1753 there were 2 methods of Chronological computation in use in the Kingdom. One known as the Civil Ecclesiastical and Legal Year commencing on the 25 March: the other known as the Common Historical Year commencing as now on the 1st of January. As different writers according to their fancy adopted either style confusion frequently arises respecting the year of any occurrence between 1st January and the 25 March following as the Civil Year was always so much behind the Historical year.

"By statute in the year 1753 it was enacted that both years should commence on the 1st of January as they have since done: but in dates previous to the enforcement of that Statute occurring during that period specified it is usual to add the Historical year as in the date of this Flood having taken place according to the Historical and more correct computation of the 20th January, 1607."[3] This confusion appears in the various accounts, making it hard to establish sequences of events, and sometimes suggests there were 2 floods.

There was widespread anger at the change, with claims that people had lost 10 days of their lives. Most accounts use the latter, modern date, but I will use 1606/7 except when quoting from sources.

'God's Warning...' opens with "Many are the dombe [doom] warnings of destruction which the Almighty God hath lately scourged upon our Kingdom; and many more are the threatening Tokens of his heavy Wrath extended towards us: all of which our bleeding hears may inforce us to put on the true Garment of Repentance and ... solicit the sweet mercies of our most loving God."

This suggests the Great Flood was not an isolated event, but arrived in the wake of other disasters, so its victims were both less numerous and less robust.

It begins with the flood approaching, then focuses in on individual accounts, constantly reminding the reader of the sense of terror, and that the victims were being punished by God, as in the Age of Noah. This probably impaired the judgement of victims, added to their panic and confusion and so delayed and reduced their ability to escape. Readers were urged to "put on the true garment of Repentance, and appeal to God's mercy". It reminds the reader of the "late grievous and most lamentable Plague of Pestilence" which killed so many people, and must also have weakened the survivors. Readers are also reminded of the Gunpowder Plot of 5 November the previous year, which "practiced the subversion of this beautiful Kingdom" and of which the aftershocks were still reverberating. It also refers to another event, without mentioning the date or place, i.e. "these late swellings of the outrageous Waters which of late happened in diverse partes of this Realme, together with the overflowing of the Seas in diverse and sundry places thereof, whose fruitful Valleys being now overwhelmed and drowned with these most unfortunate and unseasonable salt waters, do foreshow great Barrenness and Famine to ensue after it (unless the Almighty God of his infinite mercy and goodness and do prevent it)."

The size of the flood is described as "the like never in the memory of man, hath ever bin seen or heard of: The sudden terror whereof struck such an amazed fear into the hearts of all the inhabitants, of those parts, that every one prepared himself ready to entertain the last Period of his lives Destruction: Doming it altogether to be a second deluge: or an universal, punishment by Water."

This demonstrates how differently the Great Flood was perceived from modern events. To us it is extreme weather or, more recently, global warning. We have access to weather warnings, to emergency services, and often the means of flight, in cars or boats. Victims of the Great Flood could only flee on foot, or a few on horseback. There is no record of people saving themselves in boats at the outset. They sought relief in God, but this flood suggested he had abandoned them. As with the Black Death, wealth, age or piety were shown to provide no protection.

The mention of an earlier flood appears in several other accounts, which claim to compare it with the 1606/7 event. None provide dates but mention towns apparently from different regions. Added to the confusion of dates is the use of old spelling, so it is hard to establish if there were 2 separate floods. But many accounts claim 1606/7 was the biggest, a claim repeated by several local sources in Wales.

'God's Warning...' begins by listing the affected areas in the English counties of Gloucester, Somerset and Monmouth, and the Welsh counties of Glamorgan and Carmarthen, and diverse and sundry other places of South Wales. So we are dealing with the lands bordering the Bristol Channel/Severn Estuary and river, which in turn suggests the Great Flood swept in from the south west, the same direction as the Gulf Stream winds which traditionally drove ships home and still makes the region warmer and wetter than the east.

It seems the terrible day began like any other. I have modernised the language as follows:

"For upon the Tuesday being the 20$^{th}$ of January last, about 9 o'clock in the morning, the sun being most fairly and brightly spread, many of the inhabitants of those Countries before mentioned, prepared the move to their affairs, some to one business, some to an other: every man according to his calling. As the Plowmen setting forth their Cattle to their labours, the Shepherds feeding of their flocks, the Farmers overseeing of their grounds, and looking to their cattle finding therein and so everyone employed in his business as occasion required."

This suggests there was no warning, so this was no normal flood, which was usually the result of heavy rainfall.

"Then they might see and perceive afar off, as it were in the Element [sky], huge and mighty hills of water, tumbling one over another, in such sort as if the greatest mountains, in the world, had overwhelmed the lower valleys or Marshy grounds." This source seems to be on high ground, with good views of lower lands, so is unlikely to be from South Wales, the coast, or levels.

"Sometimes it so dazzled the eyes of many of the Spectators, that they imagined it had been some fog or mist, coming with great swiftness towards them: and with such a smoke, as if mountains were all on fire: and to the view of some, it seemed as if thousands of Arrows had been shot forth all at one time, which came in such swiftness, as it was verily thought, that the fowls of the air could scarce fly so fast, such was the threatening furies thereof."

Suggestions have been made that this shooting of arrows was the dazzling light at the edge of clouds ahead of a heavy rainstorm. But this is odd, as there was no rain involved.

"But as soon as the people of those counties perceived that it was the violence of the waters of the raging seas, and that they began to exceed the compass of their accustomed bounds, and making so furiously towards them. Happy were they that could make the best, and most sped away, many of them, leaving all their goods and substance, to the merciless waters, being glad to escape away with life themselves: But so violent and swift were the waves... and especially the places which lay low, were all overflown, and many hundreds of people both men women and children were then quite devoured, by these outrageous waters, such was the fury of the waves, of the seas, the one of them driving the other forwards with such force and swiftness, that it is almost incredible for any to believe the same, except such as tasted of the smart thereof, and such as beheld the same with their own eyes: Nay more, the Farmers, Husbandmen, and shepherds, might behold their goodly flocks of sheep, swimming upon the waters dead, which could by no means be recovered."

This suggests an extraordinarily high tide. The term 'devoured' rather than 'drowned' is also important, as it emphasises the violence and unnatural behaviour, with echoes of wild beasts or cannibals, which were the terror of long-distance sailors. It also suggests that the bodies were not recovered for Christian burial, causing victims to be punished both in life and in death.

"Many Gentlemen, Yeomen and others, had great losses, of Cattle, as Oxen, Kine (cows), Bullocks, Horses, Colts, Sheep, Swine. Nay not so much as their Poultry about their houses, but all were overwhelmed and drowned by these merciless Waters:"

This suggests the waters were higher than houses and trees.

"Many men that were rich in the morning when they rose out of their beds, were made poor before noon the same day: such are the judgements of the Almighty God, who is the giver of all good things, who can and will dispose of them again at all times, according to his good will and pleasure, whensoever it shall seem best unto him. Many others likewise, had their habitations or dwelling houses all carried away in a short time, and had not a place left them, so much as to shroud themselves in."

The first sentence above has often been quoted in relation to the flood, as it shows, as with the various plagues, that wealth provided no protection. That said, much of the low-lying land was home to the poor, who must have been worse hit than those on higher ground. The term 'shroud' rather than 'clothe' also adds to the terror, as it again suggests the victims were denied Christian burial, and so entrance to heaven.

"Moreover, many that had great store of corn and grain, in their barnes in the morning had not within 5 boards' space afterwards, so much left as a lock of Hay or Straw to feed their Cattle which were left. Such was the great misery they sustained by the fury of this watery Element, from which like, Good Lord I beseech him of his infinite mercy and goodness to deliver us all."

The source adds more detail, helpfully naming some of the affected towns in the region which were cited in later sources:

"All the counties along on both the sides of the river of Severn, from Gloucester to Bristol, which is about some 20 miles was all overflowed, in some places 6 miles over, in some places more, in some less. All, or the most part, of the Bridges, between Gloucester and Bristol, were forcibly carried away with the waters: besides many goodly buildings thereabouts much defaced, and many of them carried quite away: besides many other great losses of all kinds of corn and grain, and cattle that there then lost."

Loss of buildings must have been immense, especially for the poor who were more likely to live on low-lying ground near the water, in single-storey wood and mud houses which would have dissolved in the waters. In Weston–Super–Mare museum is an example of walls made of brambles and mud. But evidence of this damage has been lost, often due to the Civil War and Georgian and Victorian improvements, and more recent changes of land use and shifts in population away from poor rural areas. The mention of bridges between Gloucester and Bristol is confusing, as this suggests they were across the Severn, but it seems there was no permanent crossing until the construction of the motorway bridge between Aust and South Wales which opened in 1966. These bridges must have been smaller ones, over the many streams and drainage ditches or rheens/rhines.

The text includes details of what passed for rescue efforts: "at Aust, many passengers that are ferried over there now, are seen to be led by guides, all along the canals, where the water still remains for the space of 3 or 4 miles, or else they will be, in great danger of drowning, the Water still lying so deep there."

This is the only mention of the ferry, the survival of which seems to have bordered on miraculous. But despite this being on the coast of the estuary, there is no mention of rescue boats.

"Many dead carcasses, both there, and in many other places, of the country, are daily found floating upon the waters, and as yet cannot be known who they are, or what number of persons are drowned, by reason of the same waters, which as yet in many places remain very deep: so great was the spoil that these merciless elements there wrought and made."

This is important. The usual sources of information on deaths are parish records, with births, marriages and deaths jotted down on scraps of paper or vellum by vicars, to be added to the books in clear script at a later date. These books were held in secure trunks in churches, some of which would have been damaged by the floods, leaving gaps in the surviving records. Recording the deaths of flood victims is unlikely to have been a high priority when survival and rescue were the main focus, and dry paper was probably unavailable. Some bodies would have been washed away and never found. Some were likely eaten by starving fish and animals. Others left by receding tides were probably decayed and covered in mud so left there, nameless and forgotten. This should have led to a rise in the number of strangers' burials, but no such records can be found.

Bristol was the major inland port in the West Country; its quays were low-lying and the city surrounded by the rivers Avon and Frome. But the densely populated inner city, especially round the High Cross and Castle, was on high ground, which explains the following:

"In Bristol was much harm done, by the overflowing of the waters, but not so much as in other places. Many Cellars and warehouses, where great store of merchandise was in, (as wine, Salt, hops, Spices, and other such like Ware) were all spoiled. And the people of the town were forced to be carried in Boats, up and down the city about their business in the fair time there."

This is an oddity, as fairs tended not to be in winter for obvious reasons, as travel was more dangerous and soaked clothes were heavy and raised risks of colds etc. Also, stalls and visitors would be exposed to the elements. It seems many traders were from London, so did this provide local people with imported goods when the London merchant fleets sailed to northern Europe and the Mediterranean?

Sources claim the mayor organised boats to rescue victims from trees in the surrounding low areas to the north, showing the importance of civic authorities playing similar roles to landlords and clerics in the countryside. This was likely the inspiration for the often-reprinted woodblock images of people clinging to trees above the flood waters. It is also interesting that no mention of the city's main river, the Avon, was flooded.

"Upon the other side of the river of Severn, towards Chepstow, upon the lower grounds, was much harm done, by the violence of the water. A woman was drowned in her bed; and also a girl."

Why was the woman in bed at 9am? This suggests she was unable to escape, so was likely sick or injured, or had the woman recently given birth to the girl? It seems their deaths were exceptional. It also reflects the layout of Chepstow, with its large floodplain by the river's edge, the site of wharves, shipbuilding and ancillary trades, with the castle and the crowded old city on the heights. Those in the low region probably had warning from the rising waves so were able to reach boats or high ground, which in turn, suggests the loss of life was low. The Wye is in a narrow, winding valley, so though no records survive from further upstream, the bends likely dissipated the flood's force, limiting its surge upstream. High tides usually reach Redbrook. In the absence of any mention of heavy rain on the hills, people probably had time to flee uphill when the alarm was raised. Traffic was largely via boats, which may also have saved some lives and aided rescue and recovery in the aftermath.

"Also, all along the same coasts, up to Goldcliff, Matherne, Caldicot-Moors, Redrift [Redwick], Newport, Cardiff, Cowbridge, Swansea, Laugherne, and divers other places, of Glamorganshire, Monmouthshire, Carmarthenshire and Cardiganshire; many great harms were there done, and the waters raged so furiously and with such great vehemency, that it is supposed that in those parts, there cannot be so few persons drowned as 500: both Men, Women and Children, besides the loss of abundance of all kind of Corn and Grain: together with their Hay, and other provision which they had made for their cattle."

This suggests South and West Wales were badly affected, though the death toll seems to have been incredibly low. This author cannot find any evidence of damage west of Swansea Bay. Mention of Cardiganshire seems to provide evidence that rain was not involved, as the county includes Plynlimon, the mountain which is the source of the rivers Severn and Wye. It is terrifying to consider the outcome of the incoming spring tide meeting torrential rain and heavy runoff from the hills. Rain would also have prevented the high ground providing places of refuge, so the death toll would have been far higher and the animals and food stores would have been lost. But if Cardiganshire was affected, how did Ireland escape? The main argument that the disaster was not a tsunami is the absence of records in Ireland; the same applies to the flood.

"Moreover, there were in the places afore mentioned many thousands of Cattle, which were feeding in the Lowe Valleys, drowned and overwhelmed with the violence of the furious waters: as Kine, young beasts, Horses, Sheep, Swine, and such like, the number is deemed infinite: yea, and not so much as Turkeys, Hens, Geese, Ducks and other Poultry about their houses could once escape away, the Waves of the Sea so overwhelmed them."

The region to the east of Cardiff was at the time home to a variety of farming and food production, with fishing in the Severn, waterbirds sheltering for the winter on the wetlands, and pasture for beasts, so the flood must have damaged food supplies of farm animals and wild game for some time. Fish supplies likely recovered, but except for poultry and pigs, animals produce a single offspring per year, so the region's recovery must have relied on imports for some time.

"And that which is more strange: There are not now found only floating upon the Waters still remaining, the dead Carcasses of many Men Women and Children: But also an abundance of all kind of wild Beasts, as Foxes, Hares, Conies, Rats, Moles and such like, some of them swimming one upon another's Back, thinking to have saved themselves thereby, but all was in vain, such was the force of the Waters that overpressed them."

Though not mentioned here, there must also have been problems obtaining clean water in the aftermath. The account provides plenty of detail, though it is annoyingly vague as to the locations of individuals fortunate enough to be rescued. But they were apparently close to where the author claims to have lived so so this account is likely to be more accurate than those from further afield:

"In a place in Monmouthshire, there was a maid went to milk her kine in the morning, but before she had fully ended her business, the vehemence of the Waters increased, and so suddenly environed her about, that she could not escape thence, but was enforced to make shift up to the top of a bank to save herself, which she did with much ado, where she was constrained to abide all that day and night, until 8 of the clock in the morning in great distress, what with the coldness of the air and the waters: and what with other accidents that happened unto her, she had been likely to have perished there had not the Almighty God of his infinite mercy and goodness, preserved her, from such great perils and dangers, which were likely there to ensue unto her. But there placing her self for safeguard of her life as aforesaid, having none other refuge to fly unto, the Waters in such violent sort had so pursued her, that there was but a small distance of ground left uncovered with Waters, for her to abide upon. There she remained most pitifully lamenting the great danger of life that she was then in, expecting every minute of an hour, to be overwhelmed with those merciless Waters. But the Almighty God, who is the Creator of all good things ... sent his holy Angel to command the Waters to cease their fury: and so returned into their accustomed bounds again, whereby according to his most blessed will and pleasure she was then preserved.

"In the meantime, during the continuance of her abode there, divers of her friends practised all the means they could to recover her, but could not, the waters were of such a deepness about her, such was their want in this distress, that many perished through the want thereof.

"There was a Gentleman of worth, dwelling near unto the place where she was, who caused a goodly gelding to be saddled, and set a man upon the back of him, thinking to have fetched her away, but such were the deepness of the waters, that he durst not adventure the same, but retired.

"At last some of her friends devised a device, and tied two broad Troughs the one to the other, (such as in those counties they use to salt Bacon in) and put therein two lusty strong men, who with long Poles (stirring these troughs) (as if they had been boats) made great shift to come to her, and so by this means, through God's good help she was then saved."

Humans were not the only victims of course:

"But now (gentle Reader) mark what befell, at this time, of the strangeness of other creatures: whom the Waters had violently oppressed: for the two men which took upon them to fetch away the maid from the top of the Bank, can truly witness the same as well as her self to be true, for they beheld the same with their Eyes.

"The Bank where the maid abode all that space was all so covered over, with wild beasts and vermin, that came thither to seek for succour that she had much ado to save her self, from being harmed by them: and much ado she had to keep them from creeping upon and about her, she was not so much in danger of the Water on the one side: as she was troubled with these vermin on the other side.

"The beasts and vermin that were there were these. (Viz)

"Dogs, Cats, Moles, Foxes, Hares, Coneys, Mice and Rats. But there were in abundance that which is more strange: The one of them never once offered to annoy the other: although they were deadly enemies by Nature the one to an other: Yet in this danger of life, they not once offered to express their natural genie: But in a gentle sort, they freely enjoyed the liberty of life, which in mine opinion, was a most wonderful work in Nature.

"The counties of Glamorgan, Carmarthen and Cardigan, and many other places in South-wales, have likewise borne the heavy burden, of Gods wrath herein: And many were the lives of them that were lost through this watery destruction.

"Many there were which fled into the tops of high trees, and there were forced to abide some 3 days, some more, without any victuals at all, there suffering much cold besides many other calamities, and some of them in such short, that through so much hunger and cold, some of them fell down again out of the Trees, and so were like to perish for want of succour. Others sat in the tops of high Trees as aforesaid, beholding their wives, children and servants, swimming in the Waters. Some others whilst sitting in the tops of trees might have beheld their houses overcome with the waters. Some of their houses were carried quite away: and no sign or token left there of them.

"Many of them might see, as they stood upon the tops of high Hills, their cattle perishing, and could not tell how to succour them, and their Barnes, with all their store of corn and grain quite consumed, which was no small grief unto them.

"Many People and Cattle in divers places of these Counties, might have been saved in time, if the counties had been any thing like furnished with boats, or other provision fit for such a sudden Accident, as this was, which as God himself knows was little expected of them to have fallen so suddenly upon them. But seeing the countryside was so unfurnished with Boats much harm was done, to the utter undoing of many thousands."

The mention of extensive fisheries in the region should have provided rescue boats, but the Severn tide is so powerful and unpredictable as to make river traffic dangerous, so most people avoided it where possible. Fish were mostly caught in nets and baskets anchored near the waters' edge; they were filled by rising tides, and fishermen collected them as the tide went out.

"Some fled into the tops of Churches and Steeples to save themselves, from whence they might behold, themselves deprived as well of all their substance, as also of all their joys which they had before received in their wives and children, beware, whole Racks of Pease, Beans, Dates, and other grain were seen afar off, to float upon the Water too and fro, in the Countryside as if they had bin ships upon the Seas.

"The Foundations of many Churches and houses, were in a manner decayed, and some carried quite away, as in Cardiff, in the county of Glamorgan there was a great part of the Church next the Water side eaten down, with the Water, many houses and Gardens there, which were near the water side, were all overflowed, and much harm done."

This suggests that St Mary's in Cardiff was not the only building that had decayed for lack of funding and/or patronage in the wake of the Reformation. For many years, coffins had been washed away from the churchyard and concerns raised over the dangerous state of the church itself.

"Diverse other Churches lie hidden in the Waters, and some of them the tops are to be seen: and other some, nothing at all to be seen, but the very tops of the Steeples, and of some of them nothing at all, neither steeple nor nothing else."

The mention of church steeples underwater may be the source of the images in contemporary woodcuts, though it is hard to believe the waters could have reached so high. In poor areas, churches were similar to the houses of the poor they served. But it seems in some narrow valleys, large amounts of floodwater were forced into very narrow channels by the rising tides, suggesting these were accurate claims.

"Also many schools of young scholars, in many places of those countries, stood in great perplexity, some of them
adventuring home to their parents were drowned by the way: Other some staying behind in Churches, did climb up to the tops of Steeples, where they were very near starved to death for want of food and fire: many by the help of Boards and planks of Wood, swam to dry land and so were preserved from untimely Death."

Again, it seems unusual to be mentioning scholars as if there were large numbers of them at a time when schools were few. Most were founded by wealthy benefactors, though schools were popular in Wales under the Tudors as a means of advancement, especially in learning English. But if this happened in a major town, why isn't it named?

"Many had Boats brought them, some 10 miles, some 15, some 20, where there was never seen any Boats before." This suggests many people were involved, so why is their location not named?

"Thus God suffered many of them to escape his ireful wrath, in hope of their amendment of life: some men that were riding on the highways were overtaken with these merciless Waters, and were drowned.

"And again many have bin most strangely preserved. As for example, there was in the County of Glamorgan a man both blind and did ride and one which had not been able to stand upon his legs in ten years before, he had his poor cottage broken down by the force of the Waters, and himself, Bed and all carried into the open fields, where being ready to sinke, and at the point to seek a resting place, 2 fathoms [12 feet] deep under the Waters: his hand by chance caught hold of the Rafter of an house swimming by the fierceness of the winds, then blowing Easterly he was driven safely to the Land, and so escaped.

Also in an other place, there was a Man Child of the age of 5, or 6 years, which was kept swimming for the space of 2 hours, above the Waters, by reason that his long Coates lay spread upon the tops of the waters, and being at last, at the very point to sink: there came by chance by, (floating upon the tops of the Waters,) a fat Wether that was dead, very full of Wool: The poor distressed Child perceiving this good means of recovery, caught fast hold on the Wether's Wool, and likewise with the wind he was driven to dry land and so saved."

It is hard to overstate how extraordinary this child's survival was. Two hours in the cold water is a long time, especially if he was swimming or paddling rather than floating. Most people would have either drowned or died of exposure. Though most of these people were wearing wool, which provides good thermal insulation in the wet. England's wool trade thus probably saved many lives on this day.

"There was also in the County of Carmarthen, a young Woman, who had 4 small Children, and not one of them able to help itself. And the Mother then seeing the surges of the Waters to be so violent to seize upon her threatening the Destruction of her self and her small Children, (and as a Woman's will is ever ready in extremities) she took a long Trough, wherein she was wont to make her bread in, and therein placed her self, and her 4 Children: And so putting themselves to the mercies of the Waters, they were all by that means driven to the dry land, and by God's good providence there by they were all saved." This must have been a huge trough.

"Many more there were that through the handiworks of God were preserved from this violent death of Drowning, some on the backs of dead Cattle. Some upon Wooden planks: some by climbing of Trees, some by remaining in the tops of high Steeples and churches, other some by making of speed away with swift horses, and some by the means of Boats, sent out by their friends to succour them: but there were not so many so strangely saved, but there were as many in number as strangely drowned."

This last is an interesting statement, suggesting the death toll in the region was 50%. It seems the worst-hit region was the Somerset marshes and coastal plains, much of which is still at, or below, sea level and increasingly struggles with floods due to modern climate change. The following is a terrifying read, and a warning of what may await us in the coming years.

"The low Marshes and Fenny grounds near Barnstaple in the County of Devon were overflown, so far out, and in such outrageous sort, that the country all along to Bridge-water was greatly distressed thereby, and much hurt there done it is a most pitiful sight to behold what numbers of fat Oxen, were there Drowned: what flocks of Sheep, what heads of Kine, have there been lost, and Drowned in these outrageous Waters: there is little now remaining there to be seen, but huge Waters like to the main Ocean: The tops of churches and Steeples like to the tops of Rocks in the Sea. Great Ricks of Fodder for Cattle, are floating like Ships upon the same, through the rigour of this Element of Water: The tops of Trees, a man may behold remaining above the Waters, Upon whose branches, multitudes of all kinds of Turkeys, Hens, and other such like Poultry were fain to fly up into the Trees to save their lives, where many of them perished to death, for want of relief, not being able to fly to dry land for succour, by reason of their weakness."

"This merciless Water breaking into the Bosom of the firm Land, hath proved a fearful punishment, as well to all other living Creatures: as also to all Mankind: Which if it had not been for the merciful promise of God, at the last Dissolution of the World, by Water, by the sign of the Rainbow, which is still showed us: we might have verily believed, this time had been the very house of Christ his coming: From which Element of Water, extended towards us in this fearful manner, good Lord deliver vs all. Amen."

Reading the above, and other contemporary tracts, it is hard to understand how people coped before modern emergency services were available. The terror of what people experienced, the lack of help for many, made worse by the sense that they were being punished by God must have been terrifying and disorienting. It must have driven many to question themselves, and their faith. Just as with the Black Death, survivors must have wondered why so many were lost, and why they survived.

Is it possible that this event was the tipping point in founding colonies across the Atlantic? If God was punishing people in what they thought were their safe homes, did crossing the stormy Atlantic in a small boat seem safer than remaining? When aristocrats invested in settlements in North America, they promised plentiful fertile land and placid local people. How many survivors of this flood chose to leave home rather than endure more of these horrors they had endured?

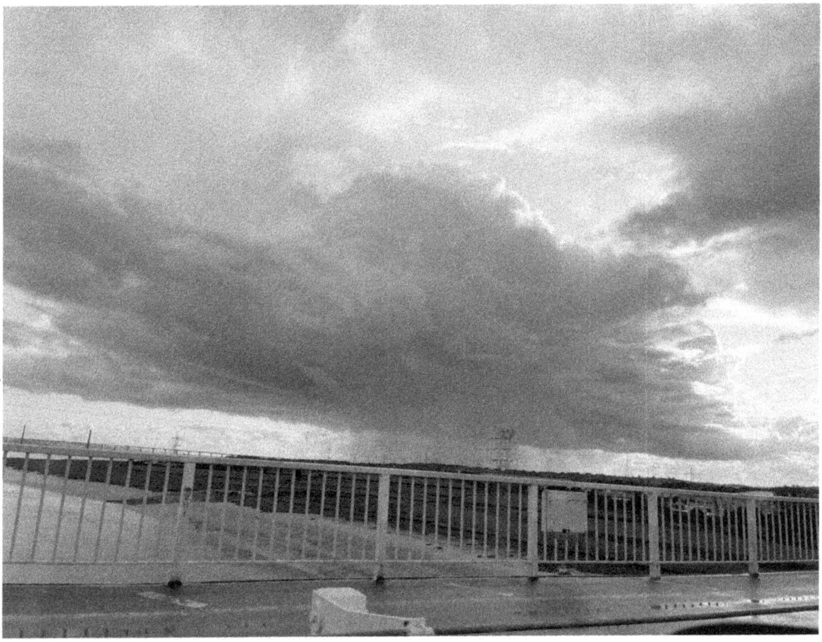

# CHAPTER 4
# LAMENTABLE NEWS

In F. J. North's book "The Evolution of the Bristol Channel', he describes the Great Flood of 1606/7 when 26 coastal parishes in South Wales were affected, extending from Portskewett in the east to Rumney in the west, so almost the whole of the coastal Wentloog and Caldicot Levels. He describes it : "the spring tide, assisted by a strong south-westerly gale, carried water over the embankment and inundated an area said to be about 24 miles by 4".[1] Thus he is another source stating that it was an extreme high tide rather than a storm. He continues: "Such inundations of low-lying tracts are due to unusually high seas breaking through or over protective barriers of natural or artificial origin: they are of temporary duration and cannot be compared with the effects of the gradual but irresistible encroachment of the sea on a subsiding shore, as was the case with the much earlier floods that are commemorated in the legends of Cardigan Bay and the North Wales coast where land, once lost, has not been regained; they do not indicate that a downward movement is still in progress."[2] This is important, as it makes a clear distinction between accounts/legends of lands beneath the sea and this single disaster, though to victims the distinction likely made no difference.

The strange term Wentloog referred to an extensive forest which survived into the early 20th century when 2,244 acres were sold by

Henry Somerset, 9th Duke of Beaufort. It lay to the north east, and partly within the boundaries of the city of Newport, and was described as the largest wood in England. It was also called Wentwood.

The extraordinary early historian of Bristol, Reverend Samuel Seyer claimed that the water came in so fast between Chepstow and the further end of Carmarthenshire that 500 were drowned. This seems a low figure for such a large, densely populated region. But the region had been depleted by famine in 1585–7 and 1593–7 and also by outbreaks of plague.[3]

Children were claimed to have been caught travelling to school, but as literacy rates were still low, this suggests a small group, or a detail added in later imprints to emphasise that even innocents were punished. Seyer also claims "many were drowned before they were aware of the danger", which emphasises the speed at which the tide travelled, and that it was not accompanied by other usual weather warnings such as strong winds, thunder and rain.

The Great Flood also needs to be seen in context. Hewlett claims there was also a Severn flood on 4 November 1636, described by others as a storm, when the sea walls were "lacerated and broken by the Hand of God". But it was not as high as in 1606/7, which raises questions as to how many other events have not survived in the records. On 1 August that year, a new stretch of the sea wall was built at Hill and more repairs were reported as needed but the Civil War intervened.[4]

Land on the river's edge was owned by different people, so its defences were not co-ordinated. Some areas were owned by absentee landlords who showed little interest beyond collecting rents, so they neglected to maintain the walls and drains. Those with enough income lived on elevated sites, such as at Almondsbury. Severn Beach, New Passage, Redwick, Northwick and Pilney which were all protected by sea walls, especially the Binn Wall, site of the New Passage, or crossing, which existed in the 16th century. Berkeley Castle was built above the surrounding plain which could be flooded in case of attack, so flooding was integral to its existence.

Sir Edward Winter was in charge of the Court of Sewers at the time of the Great Flood, and he convened a meeting to discuss the disaster on 31 March 1607, i.e. 6 weeks later. But there must have been much activity in the immediate aftermath to limit loss of lives. The Court of Sewers records are valuable because those from Monmouth, Glamorgan, Somerset and Devon are lost. It seems the court was founded following a dispute over maintenance of the sea walls at Oldbury-on-Severn which dragged out for over a century. The site was most at risk from flooding, as it had a pair of tide mills, and locals complained of the ponds storing sea water to run the mills, which raised the risk of flooding in the surrounding area.

It seems the region's land reclamation from the Severn was begun by the Romans with engineering works at Elmore, Longney, Arlingham and Hill. In later centuries there were also works at Fretherne, Saul, Frampton-on-Severn and Slimbridge. Henbury's salt marshes seem to have been drained from the 12$^{th}$ century to allow lands to be used for farming. These floodplains had built up layers of fertile silt to become valuable farmlands. But this was dependent on constant maintenance of walls and drainage ditches. This in turn required planning, co-operation and enforcement, which was possible when religious houses owned large estates. Post-Reformation, the system instead relied upon conscientious landowners to supervise and enforce maintenance, which was never a given.

The second published account of the disaster apparently comes from a broadsheet published in London. As was usual practice at the time, it was commercially published, so focused on the drama to maximise sales. But it provides much credible information, so forms the basis of later accounts of the tragedy. It states it was printed for W.W. and was sold at St Paul's churchyard, London, and the title was:

"Lamentable newes out of Monmouthshire in Wales, contayning the wonderfull and most fearefull accidents of the great overflowing of waters in the saide Countrye, drowning infinite numbers of Cattell of all Kins, such as Sheepe, Oxen, Kine, and horses, with others; together with the losse of many men, Women and Children, and the subversion of xxvi (26) parishes in January last, 1607." At the time of the flood, Monmouth was part of England, so is another hint that this was of a later date. This was followed by a crude woodcut showing people and animals in the water with churches looming above the floodwaters.

But there is a problem here, as the date of the flood was 1606 in the Old style, Julian Calendar in use from 1582, which suggests this apparently contemporary document is a year later, so raising questions as to its accuracy. Under the old system, the year started in March, so I have cited it in this book as dating from 1606/7.

The term "wonderful" is not used in the modern sense, but reflects the sense of wonder and awe which the event inspired.

It addresses the reader by begging them to note how God has punished the victims, warning that worse could befall them if they fail to mend their ways, a claim which today seems both cruel and nonsensical. This also suggests the tract was printed as a propaganda tool by evangelicals, rather than as a source of news in the modern sense.

It refers to damage caused along the Bristol Channel: "ye Sea got up between Barstaple and Bristowe as high as Bridgwater", with the whole Brent Marsh on the Somerset Levels being submerged.

The print includes the famous woodcut of people clinging to trees and waving their arms as they bobbed in the water with various beasts. A man clings to a tree, naked apart from his warm Monmouth cap. It shows the waters to be above the main roof of a church, which seems ridiculously high, but there are claims of floodwaters drowning churches. Some of the affected areas were small rural parishes, whose churches were similar to the simple single-storey cottages their parishioners lived in, so were likely lost in the flood.

Also noteworthy is that the main landlord in South Wales was the Herbert family, the dominant branch of the omnipresent Berkeley/Somerset family. They were based at Raglan Castle until it was destroyed by Commonwealth forces during the Civil War. The family became the Dukes of Beaufort in 1682 and moved to Badminton in Gloucestershire, though they still held vast estates in South Wales, where the Morgans set standards and fashions such as at Tredegar House near Newport.[5] The family's long-standing presence also helps explain the many links between South Wales and South West England, especially in the parish churches, many of which were founded by Normans with links to Bristol and Somerset.

Behind the pulpit in Nash church is a copy of a similar print, framed, with the names of the affected parishes around its edges. Some of the churches are now in private ownership or have been demolished after long-term neglect. Oddly, the letter Es have been replaced by pound (£) signs, suggesting the printing was done on the cheap.

There are several surprisingly non-local inclusions such as Bar(n)staple and Bristol, and a few which are hard to identify. But the list confirms that both sides of the Severn were affected, which adds to the problems in researching the event as each record office has its own way of collecting and recording its histories. It is also reflects the close, but now little-known, links between English and Welsh ports before road and rail travel became widespread. The church of Penarth on a hill overlooking Cardiff should have been yet another South Walian church dedicated to St Michael. But it was owned by St Augustine's Abbey in Bristol, so shares its dedication. There are also a surprising number of settlements on either side of the channel which share the same names, the most obvious of which is Redwick, probably reflecting cross channel traffic.

There is a surprising absence of evidence of rivers being affected by the flood, the Taff in central Cardiff being an exception due to the loss of its parish church.

It seems the bulk of the victims lived on the coast, surviving on a mix of farming, hunting and fishing, so their houses were "beaten down", damaged by rain and floods, causing the "scattering and dispersing the poore substance of innumerable persons". Claims have been made that the total cost was valued at £100,000, a huge amount at the time, but this seems a suspiciously rounded figure. A source has claimed that the death toll was low, but this makes no sense. There were so many ways for people to die: at the time of the disaster by being swept away by floodwaters or being trapped in houses and drowning, then in the aftermath also from hypothermia and hunger as well as shock and exhaustion.

Individual victims were noted, allowing readers to sympathise with their plight, though annoyingly, though without their names and mostly without their locations. A woman placed her 4 year old on a roof beam, hoping they would stay above the rising waters, where a chicken flew up to protect the naked child from freezing to death. This sounds like a fantasy. Mistress Van's income was claimed to be over

£100 per year, a substantial sum and her location was noted as 4 miles from the sea wall. But this did not protect her from drowning. So as in accounts of death from plague, it was a reminder that life was fragile, that wealth was no protection, and a warning to be ready whenever it might strike.

The good news is that the great landlords Lord Herbert and Sir Walter Montague were praised for sending boats and hay wagons to rescue victims. The document adds with a warning: "The Lorde of his mercie graunt, that we may learne in time to be wise unto our own health and salvation, least that these water-floods in particular prove but forerunners unto some fearfull calamities, more general."

The parallels with Noah's Flood would not have been lost on anyone, so survivors likely felt relief but also perhaps confusion and survivor's guilt.

A list of the named parishes follows, though the names of 2 are hidden by the frame. I have provided contemporary information where available. Though the majority of this list are from the South Wales Levels, there are several across the channel, such as Bristol and Barnstaple, with the former being well documented in local sources. Lambeth, Brent and Romney appear to be in London/Essex, an extraordinary claim, suggesting the flood reached the East Coast, which in turn suggests it also struck mainland Europe, which should have left documentary traces. But it seems Brent was Brean Down in Somerset. If so, this suggests the other 2 are also spelling errors. Most people lived in small clusters of houses in the countryside, too small to have names, so when swept away by this flood, they left no trace. This could have been a significant part of the population at the time, but there is no way of confirming the numbers.

The following are listed in order from west to east, with contemporary information where available. There are also several settlements named which could be beyond the levels, but I have accepted them within the region as the bulk of these sites are local. Some provide valuable contemporary details with plaques recording the highest levels the Great Flood reached reflecting their value as memorials where significant numbers of parishioners were lost. Being inside the churches, they have outlived any tombstones in the churchyards, so are invaluable records of the disaster. Dedications of churches to St Mary and St Peter generally suggest ancient foundations.

Marshfield is now part of a green corridor between Cardiff and Newport, but is still mostly rural. The Grade II* listed St Mary's seems to date from 1135 and was another site founded by Mabel FitzRobert, Countess of Gloucester after whom Cefn Mably, a region 6 miles north of Cardiff with a historic house was named. It is said to have been built in memory of her father Robert Fitzhamon, Norman ruler of Glamorgan and Wentloog. The church is one of many granted to Bristol's St Augustine's Abbey. The go-to architectural historian Nikolaus Pevsner's guide describes it as a "splendid building", by the same masons who worked on Llandaff Cathedral.[6]

Caldicot is described by Pevsner as "quite a large and handsome church, built in 3 phases from its Norman founding. Its unusual window tracery is also seen locally at Redwick & Rogiet, and the doorway similar to Redwick and Undy." Pevsner also notes "the interior has been scraped",[7] which seems to suggest that any flood markers have not survived. Given the number of buildings that were replaced or 'improved' by Victorians, this may be a major explanation for the dearth of flood evidence.

Llanfihangel is Welsh for St Michael and any churches dedicated to him are often on mountains, suggesting it would be an unlikely flood victim. It is an incredibly common attribution in Wales as the country has many hills. But the name is also applied to coastal rocks, which are also common in the region. This church is most likely on the Caldicot Levels, near Rogiet, in Monmouth. It is Grade II listed, mostly 13th century, and now in the care of Friends Of Friendless Churches.

Lanckstone is probably anglicised as Langstone, its age and/or neglect suggested by its lack of dedication. Pevsner says it stands alone south of the modern village. It is small, which adds to a sense of age, so a likely survivor of the flood. The west doorway lintel is dated 1622. Was it in need of repair, then delayed whilst funds were raised after the flood? Nearby is another small church, the 12th century St Curig, a medieval chapel converted to a stone barn, granted by its founder Robert de Chandos to his priory at Goldcliff on the nearby coast.[8]

Lambeth is unknown. Could it be a misspelling of a Welsh church which has been lost? Possibly one of the small churches shown almost covered by the waters in the woodblocks of the disaster?

In Magor, St Mary's previous name was St Bride's Brook which became the canal Mill Reen to reach the sea by a good sized creek. Abergwaitha was a major port but fell victim to a 14$^{th}$ century storm. Its remains can be seen as a surviving causeway leading to Magor. The reclaimed salt marshes show traces of Roman presence. Pevsner describes it as one of the county's most ambitious churches, dating from the 13$^{th}$ century. Both the church and its churchyard are huge. It is yet another church restored by Bristol's prolific Victorian architect and antiquarian John Norton and shows West Country elements. The village is mostly known for its motorway services, between Chepstow and Newport on the Caldicot Levels beside the Severn Estuary. It is an area known for its Roman ruins, including sea defences and/or a causeway. Many Roman artefacts have been found nearby and the village centre was in the salt marshes that were reclaimed by Romans, so it is no surprise it was hit by the 1606/7 flood. A causeway likely linked it to a now-lost harbour on the Severn Estuary. A 13$^{th}$-century boat was found buried there in 1994 close to Magor Pill, carrying Glamorgan iron ore. Magor Marsh is now a 90-acre Gwent Wildlife Trust nature reserve, crossed by many reens and is rich in wildlife.

Undy is nearby, described by Pevsner as "a small church but a puzzle, the oldest part dating from the 13$^{th}$ century, and another Norton restoration". It lies north of Magor Reen so was likely flooded, and is now largely a commuter village

Llanwerne – could this be the misspelled — so untraceable — flood victim of Lambeth? This area is now dominated by the huge Spencer Steelworks, opened in 1962. But as if from another world, beside this beast is Lord Rhondda's seat of Llanwerne Park. The Church of St Mary seems to be partly pre-Norman Conquest. It is east of the village and described as "small and isolated". Its north wall is windowless, suggesting it was built during a very cold period, which seems to explain the sealing of many north doors to keep out Arctic winds. Nothing to do with witches.

Matherne is a historic village and conservation area about 3 miles south west of Chepstow near the Severn Estuary and Bristol Channel, but it is now divided by the M48 motorway. It apparently originated when St Pierre Pill, off the Severn Estuary, was bigger and more important, meeting an ancient ridgeway through Shirenewton to Monmouth. This inlet became named Portskewett after another nearby village, which is listed as a flood victim. It was named after the burial place of a martyr, and largely rebuilt by the Bishop of Landaff — It seems significant that this took so long — in 1601–17. Was this delay due to the Great Flood? Its church is tiny and Grade 1 listed.

Milton is the name of several roads, and a hill in Llanwern, another settlement listed which seems an unlikely flood area. But it is near several reens and north of Broadstreet Common and the known flood victims of Goldcliff and Whitson.

Nash St Mary is described by Pevsner as a surprisingly large church, rising above "a scattering" of council houses, with the air of a rundown seaside village. It has a tall steeple, built by Eton College, which held the rectory from 1450. Its Georgian nave is large and still filled with Georgian box pews with a minstrel gallery at the west end. Its tall 15th century steeple is unique in Wales, so apparently survived the flood.[9] Wikipedia calls it "the Cathedral of the Moors", a term also shared with more justification by St Peter. It is Grade I listed, mostly due to its "medieval tower with fine spire". It originally belonged to Goldcliffe Priory, but was largely rebuilt in the 16th century, suggesting it was neglected before the Reformation.

Peterston, St Peter is called by Pevsner "the queen of the churches" on the levels, and "the noblest and most beautiful Perpendicular church in the whole county".[10] Its "ambitious West Country character" reflects its construction by St Augustine's Priory, Bristol. The saint is shown with the keys to heaven so this was the most important and best-funded church in the region. Like nearby Marshfield, it was founded in 1142 by Mabel Fitzrobert. The building is an extraordinary discovery; a large, impressive church but now closed to the public with signs warning of dogs running loose. Cadw listed it as early as 1963 as Grade I: "a mid 15th –century Perpendicular style, restored after The Great Flood, with a marker for the water level, and again in 1887 by George F.R. Walker, Bart in memory of his wife, Fanny Henrietta, 3rd daughter of Baron Tredegar" so had links with nearby Tredegar House.

Cadw adds "It is said to have monastic origins". Much of the 15th - century church seems to survive. The parapet is "highly decorative West Country style", typical of the Devon area. The "sophisticatedly decorated tower" is similar to that of nearby St Brides. The interior includes a very rare rood stair and a 15th century polygonal font. The grounds for its listing are "a rare, surviving example of a large Gwent Perpendicular church with a fine interior".[11] But at the time of the sale, its wooden pews were rotten and slates were missing from its roof, so as with Goldcliffe, which was also sold, it suggests the parishes were too shrunken to continue. No wonder it has dogs roaming the site.

Savills listed it for sale in 2021 as set in an acre of gardens with impressive views of the estuary for £410,000 with a 999-year lease. At the time, it was part converted for domestic use, with potential for further changes.

Portescuet, or Portskewett, St Mary is described by Pevsner as "a small Norman church of local limestone, with a surviving 12th century chancel arch".[12] Ominously, it had a reordering in 1925, and the only monument mentioned is from 1819, and when I visited I saw no sign of any flood damage.

Llanerstone cannot be traced.

Redwick(e), Newport is the most famous and in many ways the most fascinating survivor of the Great Flood. It is described by Pevsner as "one of the finest churches on the Levels".[13] Its parapet is claimed to be similar to Peterstone, the interior rebuilt and tower remodelled in the 14th century great age of church buildings, then repaired in the 19th century by Norton of Bristol who also built the full immersion font. It is near the coast, east of Goldcliff. Wikipedia describes it as "on the flat coastal lands reclaimed from the Severn Estuary and Bristol Channel, part of the Caldicot and Wentloog Levels".

Redwick is unusual in many respects, including having been owned by Tintern, and oddly, has had several dedications, and is now returned to its original, that of St Thomas the Martyr, the original doubting Thomas. This is an unusual dedication as he was martyred in Mylapore near Madras in 72 AD, the port of which was destroyed in 1341 by a huge flood which realigned the coast. Does this suggest the foundation here was linked to major storm and/or flood damage? Is this another example of the Flemings arriving, or is this an earlier, Norman event? Was the South Walian coastline realigned at this time?

The church is also unusual in being Grade I listed and second in size in the area only to St Peter. Its pre-Reformation bells were made in Bristol, again reflecting the close ties and of water-bound trade with England. Nearby, 4 timber buildings have been recovered from the foreshore and others from intertidal peat which were likely used in summer pastures, echoing the practices in Somerset. The porch has a mark showing the height of the 1606/7 flood, but as this is a rebuild, the 5-feet height may not be accurate. Another flood mark, but lower, is on the buttress near the main gate. The Bible of buildings history, Pevsner fails to mention these or the event.

Rogiet could refer to several settlements, most likely to the west of Caldicot.

Romney was probably Rumney, a settlement now absorbed by Cardiff, as a river of the name runs through Caerphilly to reach the coast on the eastern edge of Cardiff, passing Rumney Hill Gardens, likely the remains of a large house. The river was the original boundary between England and Wales.

Newtone: Was this Newport?

Bristowe and Barstaple provide rare sightings across the Severn

W(h)iston does not rate a mention in Pevsner, and its church is now redundant, used as storage by local farm.

Wilfrick : This seems to be Wilcrick, near Bishton, West Moors. It was dedicated to St Mary, so is another ancient site. Pevsner notes it at the base of Wilcrick Hill, with a stone dated 1621 placed on its east wall.

St Pierre: This is south of Mathern not far from the Severn. It was a port, linked to the Severn by a tidal pill, but is now a golf and country club and hotel which makes use of the lakes.

St Mellons is now a leafy suburb, devoured by the west of Cardiff.

St Bride's, Netherwent is 2 miles north of Magor, 3 miles west of Caerwent in a village of 13$^{th}$ or 14$^{th}$ -century foundation, formerly surrounded by about 10 houses before the village was largely abandoned in the 18$^{th}$ century. The church suffers badly from damp, a victim of the Severn Estuary's 'Flat Moors' into which it has long been sinking. It has a flood marker in the porch, but is seldom open and with declining aging congregation, its future is again in doubt.

Ifton is probably a suburb of Newport.

Gold(en)cliff(e): A small church to the west of Whitson, its single room is the remains of a Benedictine priory; its monks were from Bec, Normandy. It has a plaque in the north wall of the chancel, near Roman remains of a sea wall, suggesting links with their base at Caerleon.

Christchurch is an odd inclusion, as it seems to refer to the large church on the heights above Newport, making it an unlikely flood victim. It was painted white as it served as a landmark for shipping. To the north the land slopes down to the River Usk and the town of Caerleon, suggesting parishioners may have been flooded by the river and sought shelter there.

Caldicot is between the M4 and M48 near the levels

Bashallecke is probably Bassaleg on the west of Newport where many of the Morgan family of Tredegar House are buried.

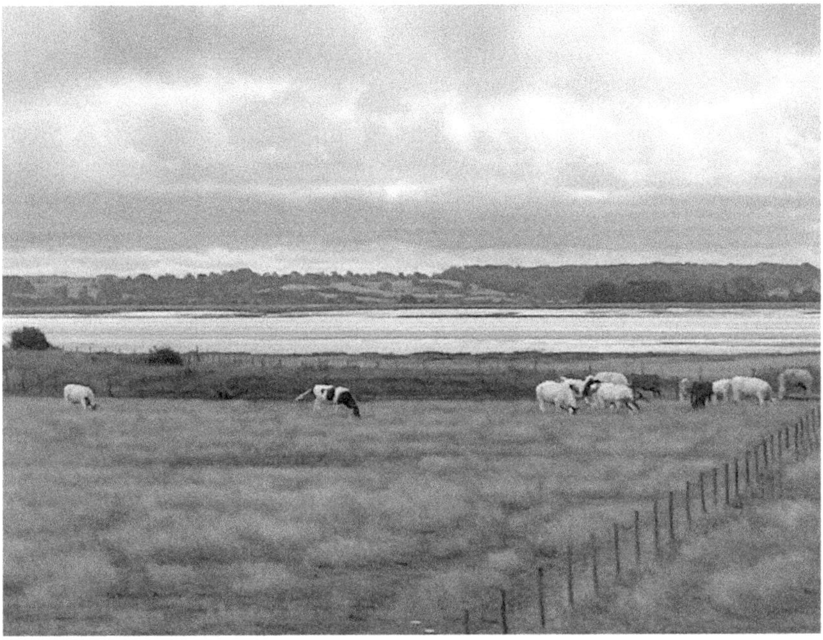

# CHAPTER 5
# SEYER'S MEMOIRS

Bristol has been a town and county since 1372; its history is well recorded and it was near the centre of the region affected by the Great Flood. It has one of the highest tidal swings in the world, which was strong enough to drive small ships to and from the port along the meandering River Avon. But as ships became bigger, they had to be towed the final distance, often by pilots from Pill.

'Memoirs Historical and Topographical of Bristol and its Neighbourhood' by cleric and schoolmaster Samuel Seyer (1757-1831) provides a lively, well-written account of the Great Flood, though annoyingly cites only a single original source towards the end. It also places the 1606/7 flood in context, as the most spectacular and destructive in an age of extremely unsettled weather. Food shortages also weakened people, raising the threat of sickness, especially the plague. "The 18th of July, 1603, the great plague began in this city, in Pepper-alley in Marsh-street, and continued 18 months: whereof died between July and Michaelmas following, 100 persons or thereabout. All that time and the most part of the year 1604, the sickness was very hot and did not wholly cease until the beginning of 1605."[1]

This was only a year before the Great Flood struck, so the population of the city and its surrounds would have been less numerous and less healthy, so in a poor state when the flood arrived. But there was more suffering to come.

"4 October 1604 was the greatest snow that was ever by the memory of man, which continued for days and by reason that the leaves were on the trees, very many were thrown down by the roots and many others were broken in pieces." This must have killed many plants and beasts, including humans both directly and via food shortages, so people were further weakened before the famous flood hit. Loss of trees would have harmed a range of local industries, especially house construction and shipbuilding, plus fuel for heating, cooking and businesses, though coal mines were expanding nearby in Somerset.[2]

The following fills in some gaps in previous chapters. It is also valuable in confirming the date of the disaster, i.e. 1606 Old Style, 1607 modern:

"20 of January 1606-7, being Tuesday in the morning, the wind blowing hard at south-west, there was so great a flood at high-water, that the sea broke over the banks, and overflowed all the marsh country in England and Wales, drowning their cattle, and carrying away their corn and hay, some houses and many trees. Some lost their lives, and many saved themselves by climbing trees and mows. In the marsh country about Aust and Henbury, the flood was so high that it could not all run off again, but remained a fathom (6 foot) deep and the people on the trees could not come down, but remained there 2 or 3 days."[3]

This confirms the flooding was due to a high tide overtopping the defensive banks, and the depth of 6 feet, aka a fathom, has been cited at other affected sites. This also suggests that a fathom was the standard height of the sea walls or banks at the time. This use of "all the marsh country" poses difficulties in assessing the extent of the disaster, as E. E. Baker mentions places on England's East Coast which he claims were flooded in 1606/7. Also, the regions flooded are hard to locate as there has been extensive land reclamation and drainage since then.

When the waters rose, the city responded promptly. "The Mayor,

Mr Barker, hearing thereof commanded cock-boats to be hauled thither to fetch them off, that they might not perish." These were small boats used mostly for conveying goods and people to and from large ships to landings etc. "In the city, it rose on the Back [Welsh Back, where market boats supplied the city with fresh food] 4 ½ feet above the street: so that a small boat about 5 tons came up laden to St Nicholas crowd [crypt] door; and the boatman put his hook against the lower step and thrust off his boat again. All the lower part of the city was covered; it was in every house on the Back and most part of the Key, [now The Centre] doing much hurt in cellars to woade [blue dye for cloth], sugars and salt; butts of secks [Sekt wine] swam in the cellars above ground, was therefore worse in vaults under ground. In Redcliffe, Temple and St Thomas streets, [the southern parishes, formerly in Somerset] the water was so high as men's girdles. In St Stephen's, St Thomas and Temple churches, it was half way up the seats. The Bridge was stopped and the water bayed back higher towards Redcliffe-street."[4]

The medieval bridge was of 4 arches, solidly built but with arches too narrow for most boats to pass.

"It rose 5 feet at Trin-mills. At its return it brought great trees down the river Avon, but did no harm to the bridge."

Trin Mills was a dual water and tide mill opposite the popular Ostrich Inn which survives on the modern Floating Harbour. It was driven by the Malago river from Bedminster south of the city where it was notoriously prone to flooding. The low-lying region was linked to the main city by a causeway which also flooded. It seems likely that much of Bedminster was thus under water.[5]

Further details of the mischief done by this calamity may be seen in 'The Gentleman's Magazine' for July 1762; the writer says it was extracted from an unnamed pamphlet written soon after the event and preserved in the Harleian Library. Confusingly, it is dated Tuesday Jan 27, 1607, but was this a week, or a year and a week after the incident?

He describes "the first bursting of the sea over the banks in prodigious high waves" which were tremendous. "The whole vale from Bristol to Gloucester was overflowed 6 miles distant from the river on both sides. Most of the bridges were destroyed." He seems to be citing the same source as Jones, but he also fails to name it. Strangely, there is also no mention of the River Avon, suggesting the force of it had dissipated before it reached the city.

"From Chepstow to the further end of Carmarthenshire it came so fast, estimated 500 persons in moderate computation, lost their lives besides many thousands of cattle, vast numbers of sheep, hogs, horses and poultry and many climbed to the top of houses, trees and towers where they saw their cattle and other substance perish and sometimes their wives and children without being able to afford them any assistance.[6]

"At Cardiff a great part of the church [St Mary] next to the river was carried away.

"Children at school, and travelling on the road were alike involved in this general calamity. If they fled to the house tops or tops of hills they were alike in danger of perishing by hunger and cold. But many were involved before they were aware of their danger.

"On the English coast the same calamity was suffered, the whole way from Barnstaple to Bridgewater it is now the most pitiable site to behold. What numbers of fat oxen were drowned, what flocks of sheep, what herds of kine were lost. There is now but little to be seen but huge waters like to the main ocean: the tops of churches and steeples like the tops of rocks in the sea: great ricks of fodder for cattle are floating like ships upon the waters and dead beasts swim thereon, now past feeding. On the same tops of trees, a man may behold remaining about the waters in whose branches multitudes of turkies hens and other poultry were fain to fly up to save their lives, where many of them perished for want of relief, not able to fly to dry land by reason of their weakness."[7]

The victims must have still been struggling to recover from the Great Flood when less than a year later, the region was again struck by extreme weather as Seyer continues:

"In 1607, on Nov 20th, began a very severe frost, which lasted 9 weeks and more to the 18th February. The Severn and Wye were frozen from Bristol to Gloucester and upwards, so that people did pass on foot from one side to the other, and played gambols, and made fires to roast meat on the ice: during which time no trows nor woodbushes could come to the city. Yet in all this time the river at the Back and Key and so downwards to Hungroad was not frozen, as it is reported to have been anno 1564."[8]

These variations in the freezing of the rivers seems curious, but the Severn is littered with shoals, so there would have been lots of shallow, non-moving water which froze easily. All these examples of extreme cold suggest there was a failure of the Gulf Stream which brings warm air from the Caribbean.

Bristol's exemption from this deep freeze seems seems odd. But it was probably protected by the warmth generated by organic waste and the constant movement of shipping in and around Bristol and the many tidal creeks.

"When the frost broke, there came swimming with the current of the tide huge flags [stones]of ice, which did endamage [sic] many ships, that were coming up the channel into Hungroad."

Again, this seems odd, but this referred to the old, medieval bridge, with its thick piers and narrow channels for the water, which meant it was easily obstructed. The population upstream was lower, so the river would be cleaner and vessels much smaller, creating fewer currents to prevent ice forming.

The repercussions for the public, especially the poor were dire. "The frost made corn very dear, and killed that which was in the earth, so that the summer following was very scarce and dear: and it starved most of the fowls of the air, especially the blackbirds and thrushes; that in riding 100 miles in the summer following a man could scarce see a blackbird." This feels very end of days.[9]

Latimer adds that the harvest of 1607, i.e., following the Great Flood, was "extremely disastrous" so the Corporation took emergency measures to prevent famine. They started by taking a census to establish how much corn was needed. They ordered wheat from Milford Haven and Ireland, suggesting these places had not suffered the Great Flood. By April 1608, 1,000 bushels of wheat were imported and the extraordinary sum of £1,000 was allocated to purchase corn from Holland, suggesting how widespread the shortage of food had become, but also that the extreme weather had not affected the East Coast and North Sea. In the summer of 1610 food prices soared due to a drought and caused much distress among the poor.[10]

In case there is any doubt about how extreme the 1606/7 flood was, and how long it remained in cultural memory, Seyer added:

"On Friday night the 12th November 1636 about 9 o'clock, the southwest wind blowing very hard upon a full spring tide, caused so great an inundation of waters to flow, that all the shops and cellars on the back and key were filed therewith and received much loss by that sudden and unexpected storm: the rage whereof brake in over the sea walls in and about Kingstone, Cleeveden, the low marish grounds between Bristol and Aust, and other low grounds both in England and Wales, doing much harm, and drowning many sheep and cattle. Yet this flood in my judgement and by the judgement of others, who set marks for both, was not so high by a foot, as it was in the last great flood in January 1606/7."

This reads like the same extreme weather as in 1606/7: no rain is mentioned in either, only wind and especially high floods. The mention of flood marks is interesting, suggesting this was more common than surviving records suggest, but I have found no mention of any markers for this later storm. Has it become confused with that of 1606/7?

Seyer reports yet another disaster from 1696, this time completely out of season when "In the month of July were uncommon rains, which continued without ceasing from Thursday the 9th in the afternoon, until Sunday morning, causing so great a fresh [freshwater floodtide] in the rivers here, that great quantities of hay were carried down and lost." Curiously, there are no mentions of fatalities.[11]

A very odd event was recorded on 9 September 1707 when "the moon being 9 days past the full, it was observed that the first tide of flood which according to its usual custom should have been at Bristol about 11 o'clock, came in about 8 and flowed about a foot at the Gibb [on the corner of King Street near what is now the city centre] and then ebbed; and afterwards on the same morning it came in again at its customary time, viz., about 11, and flowed as usual; so that it flowed and ebbed twice within 12 hours."[12]

This is one for the astronomers rather than for this climate record. I cannot find any record of it elsewhere.

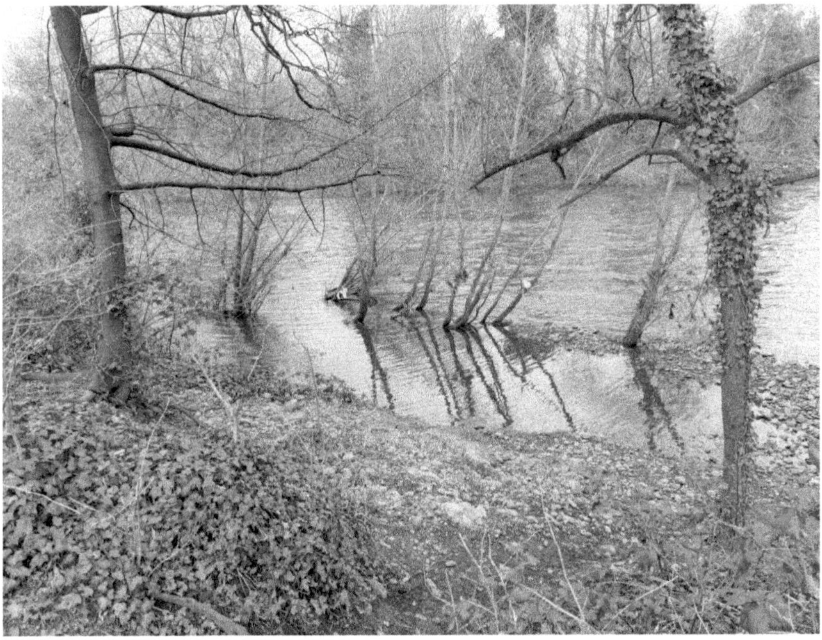

## CHAPTER 6
## BAKER'S ACCOUNT

An extraordinarily detailed account of the 1606/7 event was published by Ernest E Baker in 1884 at the Gazette office, Weston-super-Mare, south of Bristol. At the very least, it shows that the Great Flood was still remembered, but this small booklet also continues the tradition of citing the extreme weather event as evidence of God's wrath, punishing his people for their sins, so was part of the Victorian evangelical movement.

The title on the cover is 'Floods in England, 1607'. But again, which year was this? The old dating system or the modern one? The copy in Bristol Reference Library was sent by Baker to Mr Nicholls at the city's free Library, but a note adds that the recipient had died 2 years before.

The title page promises "A True Report of Certain Wonderful Overflowings of Waters in Somerset, Norfolk, & other parts of England AD 1607" in a rich salad of fonts. It is full of valuable details, but annoyingly reads more like a scrap book, with no attempt to create a united narrative.

This is followed by a preface by a William Bower: "As the interest taken in County histories and topographical works generally has increased so surprisingly during recent years, and is still increasing, few if any explanations are needed for placing before the public a reprint of a very rare and exceedingly interesting black letter [Old English or Gothic fonts] tract of 1607 relating to the great floods and inundations in Somersetshire and five other counties in England in that year." All previous accounts list Monmouthshire, Gloucester, Somerset and Devon, so the inclusion of Norfolk makes this work confusing from the outset.

"There can be little doubt that the account of these floods, and of the great loss both of life and property caused thereby, contained in this tract is truthful, and is not in anywise exaggerated, for there is no lack of corroborative evidence, and evidence too of the best sort, records made when the events occurred."

He provides 2 sources as evidence, firstly the famous board in the porch of Kingston Seymour's parish church (which he claims is called "Kingstone Seamore", showing its ownership by the Seymour family had by then been forgotten) with the following text: "January 20 1607 & 4th of James 1. An inundation of the sea-water by overflowing & breaking down of the sea banks: happened in this Parish of Kingstone-Seamore, and many others adjoining: by reason whereof many Persons were drown'd and much cattle & goods were lost; the water in the Church was 5 feet high & the greatest part lay on the ground about 10 days."

This record again suggests that neglect of coastal defences played an important role in the flood. The church is in a loop of a stream, with no protective banks so was — and is still is — subject to frequent flooding. How many other churches and settlements on the levels on both sides of the Severn were likewise at high risk every winter?

Baker continues: "Secondly, we find more particulars of "Theise late swellinges of the outragious waters" in another black letter tract of 1607 entitled "God's Warning to his People of England by the great overflowing of the Waters or Floodes, lately hapned in South Wales, and many other places. Wherein is described the great Losses, and wonderful Damages, that hapned thereby; by the Drowning of many Townes and Villages, to the utter Undooing of many Thousandes of People."

'God's Warning...' was dealt with in chapter 4, but the author adds :

"We think that every place mentioned can without much trouble be identified with some existing village or town notwithstanding that the spelling, in some cases, varies considerably."

This suggests that the countryside down to the smallest settlements recovered from the flood. But the same source claimed that some villages were devoured by the storm, which is confusing. It seems some were washed away, whilst others have been absorbed into other settlements where their names survive.

Though dated 1607, this title covers a much wider range of land affected. "A true report of certaine wonderfull overflowings of Waters, now lately in Summerset-shire, Norfolke, and with other places of England: destroying many thousands of men women, and children overthrowing and beating downe whole towns and villages, and drowning infinite numbers of sheepe and other Cattle."

It is illustrated by a woodblock image similar to those of earlier publications, with people and animals struggling in the water, people clinging for their lives to trees, and waters up to the roof of a church. It was printed at London by "W.I. at the sign of the Gunne at the north door of Paules", so again was published a long way from the disaster.

But this is odd, and differs significantly from other accounts and modern research which refer to the flood as affecting South Wales and the West of England. Was the flood big enough to reach counties on the opposite sides of England? If so, why is he only citing Norfolk? What are the odds of both sides of England being drowned at the same time, apparently leaving Ireland unharmed? If this was not the case, it raises questions as to the accuracy of Baker's text from the outset.

This is followed by an address headed "TO THE READER" with "I have to these late accidents [whereby some parts of this our kingdom have been punished], added an account of a similar disaster from the year 1570." It seems the aim of the booklet was to compare one with the other so that "God's Justice and mercy may both be seen: If those Waters of his wrath [poured down then] were more cruel than these."

This sounds like a reasonable piece of historical journalism, but then he turns up the Old Testament vibes with "It is a sign and a comfort let it be unto us that he doth but still threaten & shake the rod, for no doubt but our faults at this time are as great as in those days: If this affliction laid upon our Countrey now, bee sharper than that before, make use of it; tremble, be forewarned, Amend, least a more fearful punishment, and a longer whip of correction draw blood of us. Farewell."

As with earlier accounts, this is an evangelical, as well as a journalistic tract where the lines between the 2 are often blurred. Also, it reads more like a book of press cuttings rather than a joined-up history. But it has some fascinating detail on how people responded to floods and how these disasters were dealt with. Given the dearth of surviving accounts, it is worth diving into. But as with earlier accounts, it is often hard to establish the time and especially the places referred to.

The sites affected are different to those in earlier chapters. He refers to Somerset, omitting Wales and Gloucestershire, but oddly includes Norfolk and other parts of eastern England. He seems to be claiming the 1606/7 flood affected both sides of England, which generally have very different weather systems, and this makes no sense if he was referring to flooding. The west coast is fortunate to benefit from the Gulf Stream, a current which brings warm weather and rain from the Caribbean, whilst the East Coast is colder and drier, sharing its weather with the western coasts of northern Europe.

The date of the flood is given as 1607 in this booklet. But it is unclear which dating system this refers to. Was this the Old Style, Julian system, or the new, Gregorian, which came into use in Europe from 1582 but was not adopted in Britain and its colonies until 1752? If this book is using different sources, are we dealing with a flood in the west of southern Britain which was followed by another in the east a year later? This confusion continues in our modern age, as contemporary flood markers on the South Wales Levels date the storm/flood to 1606 but a marker recently placed on an outside buttress uses the modern dating system of 1607.

So Mr Baker seems to be an odd mix of antiquarian/historian/evangelist warning of God's revenge on those who have strayed from the path of righteousness.

But what is the significance of the year 1570? Wikipedia claims this was the All Saints' Flood of 1 and 2 November, described as the most extreme event of the Little Ice Age, which struck the coasts of Calais, the Netherlands, Eastern Frisia and Norway, drowning an estimated 25,000 people. The water rose more than 4 metres above mean high water, well above the existing dykes. Near Antwerp, 4 villages disappeared beneath the mud, and an island was lost off the coast of Zeeland. An estimated 80% of the land was inundated and the high-tide mark in a church near Emden showed an astoundingly high 4.4 metres. Many sea defences were breached and the salt water caused long-term damage to farms. Between the rivers Ems and Weser an estimated 1,000 people drowned, and many more were made homeless, as whole villages were destroyed. The remnants of Doggerland between south west Denmark and Flanders were swept away, and North Sea defences in the East of England were lost after decades of land reclamation. These walls, drainage ditches, gates and windmills had allowed the expansion of farming to support growing populations.

It was one of several dramatic floods which hit coastal England. Mumby Chapel's village and church were destroyed.

As in England, in 1606/7 the Netherlanders saw the flood as God's punishment. But unlike in England, a specific reason was provided: it was seen as punishment for not defeating their Catholic Spaniard rulers. Baker claims the flood inspired further uprisings against the Spanish after a failed attempt in 1568. In 1572, the badly flooded regions of Zeeland and Holland were liberated, which shows how resilient the people of Lowland Europe were.

Baker's 'Newes Out of Summerset shire' begins:

"Albeit that these swellings up and overflowings of Waters proceed from natural causes, yet are they the very diseases and monstrous births of nature, sent into the world to terrify it, and to put it in mind, that the great God [who holds storms in the prison of the clouds at his pleasure, and can enlarge them to breed disorder on the earth when he grows angry] can as well now drown all mankind as he did at the first: But yet by these gentle [?] warnings, he would rather have us come unto him, and fly from the points of more deadly arrows of vengeance, than utterly to perish. He fills out the measure of his chastisement according to the quality and proportion of our offences: for as the waters transgress and break their bounds, to the destruction of the fruits of the earth and to the taking away of the lives of man and beast.

"So have we that should be subjects to the Almighty King, and [by our oath of Christianity] ought and are bound to paye fealty & allegiance to our Lord and Master, gone beyond the bankes and limits of all obedience, ye taking away of his love, without which we cannot live, and to the unrecoverable undoing of our own selves."

This is followed by a preface which explains his reason for publishing the booklet.

"As the interest taken in county histories and topographical works generally has increased so surprisingly during recent years, and is still increasing, few if any explanations are needed for placing before the public a reprint of a very rare and exceedingly interesting black letter [Gothic typeface designed to force slow, deep reading] tract of 1607 relating to the great floods and inundations in Somersetshire and five other counties in England at that year."

It seems he is referring to the modern, Gregorian calendar, when the year begins in January, rather than the older dating system with the Julian new year beginning on 25 March. This means the Great Flood struck in 1606 on the Old (Julian) rather than in 1607 on the new (Gregorian) calendar, which continues to cause problems with records of the event. For accuracy, it should be recorded as the year 1606/7.

"So have we that should be subjects to the Almighty King, and [by our oath of Christianity] ought and are bound to pay fealty & allegiance to our Lord and Master, gone beyond the banks and limits of all obedience, to the taking away of his love, without which we cannot live, and to the unrecoverable undoing of our own selves.

"Sin overflows our souls: the seas of all strange impieties have rushed in upon us: we are covered with the waves of abomination and uncleanness: we are drowned in the black puddles of hellish iniquity: we swim up to the throats, nay even above the chins in covetousness, in extortion, in sensuality, in every one against the other, in contempt of our Magistrates, in neglect of our laws and in violation of those divine statutes, the breach of which is a condemnation to death, and everlasting living in Hell's fire.

"Many a time have we been summoned to an account for these

riotous abuses and misspending the talents put into our hands, we have shifted it off with counterfeit sorrow for what we have done, and with promises to become faithful servants, and new me: yet grow we worse at noon than we were at the sun's rising, and at his going down he "blusheth to behold us in our nautines".

"To a strict and strange audit therefore doth God not only call some of our countrymen now on a sudden but also to frighten us the more to make us look about, doth he strike our cattle with diseases: he takes away the lives of our beasts fit for labour; he destroys the cornfields, and threatens us with famine: he undermines our houses with tempests, to make fear and desolation.

"Read, therefore, and read with trembling these his late dreadful judgements, mock not our selves with vain hopes, but know that if earthly fathers may be drawn away to forget their own children, our heavenly father may by the vileness of our souls be drawn to shake off his own people. Listen then how he menaces, and stand amazed at the wonders of his wrath."

This is extraordinary language, a precursor of the most extreme of modern televangelists in the USA, and perhaps explains why only 200 copies were printed. Baker was writing in the Industrial Age, when scientists, chemists and engineers were creating much of what we recognise as the modern world, so he was likely a small voice shouting into a storm of indifference. Yet this pamphlet contains valuable records that do not survive elsewhere. He begins with the famous flood of 1606/7:

"In January last towards the end of the month [why so vague?] the sea at a flowing water meeting with Land-floods, strove so violently together, that bearing down all things that were builded to withstand and hinder the force of them, the banks were eaten through and a rupture made into *Somerset-shire*."

This reads like a spring tide, as claimed by most sources, an event which creates the famous Severn Bore, often at the spring and autumn equinoxes, so this disaster was unusual not only for its size, but for its timing.

"No sooner was this furious invader entered, but he got up high into the land, and encountering with the river *Severn*, they both boiled in such pride that many miles, [to the quantity of 20 miles in length, and 4 or 5 at least in breadth] were in a short time swallowed up in this torrent. This inundation began in the morning, & within few hours after, covered the face of ye earth thereabouts [that lay within the distance before named] to the depth of 11 or 12 foot in some places, in others more."

This is significantly higher than the fathom of water widely cited in South Wales and Bristol.

"The danger that this terrible tempest [the first sighting of this term] brought with it wrought much fear in the hearts of all that stood within the reach of it, but the sudden and strange cruelty of it, bred the greater terror and amazement. Men that were going to their labours were compelled [seeing so dreadful an enemy approaching] to fly back to their houses, yet before they could enter, death stood at the doors ready to receive them. In a short time did whole villages stand like islands [compassed round with waters] and in a more short time were those islands undiscoverable, and no where to be found. The tops of trees and houses only appeared [especially here where the country lay low] as if at the beginning of the world towns had been built in the bottom of the sea, and that people had played the husbandmen under the waters."

This echoes the legend of Atlantis, and also stories of sunken forests told by Gerald of Wales and others. It also raises questions as to how many settlements, too small to warrant being named on maps, were completely lost.

"Who would not have thought this had been a second *Deluge!* for at one time these inhabited paces were sunk clean out of sight. *Huntsfielde* [a Market Town in the said Shire] was quite drowned."

This name cannot be located, but it must have been a significant place to hold a market licence. It is probably the modern Huntspill, suggesting the name change reflected its altered geography, from a field to 'pill', or tidal stream. South of Burnham-on-Sea, though further inland, it is now surrounded by rhines for drainage. Wikipedia describes it as a former civil parish, which also fits. More significant is its listing in the Domesday Book as "Huna's Pill", the latter showing it was a tidal inlet and harbour. It is near the Huntspill River, the mouth of which silted up in medieval times, and was flooded in the 1606/7 disaster. Its church was gutted by fire in 1878 which likely destroyed records of it.

"*Grantham* a village utterly over-flowne." This is a confusing inclusion here as the only place of that name is in Lincolnshire. Was it destroyed by the flood, its location long lost? Or has Baker got his notes confused?

"*Kenhouse* another village covered all over." The only trace here is a Grade II listed building in Frome.

"*Kingston* a thyrd village likewise lies buried in salt water." This must be the village of Kingston Seymour, as mentioned previously.

"So [besides other small cottages standing in valleys] is *Brian Downe* a village quite consumed." This must be Brean Down, now a holiday village which has been largely absorbed by its neighbour Burnham-on-Sea. The term 'down' seems misleading, but it derives from the Celtic term 'dun', meaning hill (fort), as in the large open space in Bristol's northern suburb which has spectacular views of the Avon Gorge.

Wikipedia describes Brean as a headland rising 318 feet high, stretching 1.5 miles into the Bristol Channel, part of the line of limestone which extends from the Mendips into the sea for over a mile to the islands of Steep Holm and Flat Holm. Not surprisingly, it seems it was sparsely inhabited at the time of the flood. Brean Down is part of the parish; its highest point is 321 feet above sea level and likely provided refuge for victims when they fled from the coast.

Mee provides more detail describing it by the mid-20th century as having expanded into "a straggling collection of bungalows with a long narrow green jutting a mile into the sea, but 200 foot above, overgrown with grass and wildflowers, is a sanctuary for wild birds. There is another sanctuary on Steep Holm, and beyond that also at Flat Holm."[1]

Its parish church of St Bridget is a rare dedication. But as with the 3 across the Severn, this suggests it was built to Christianise locals. Pevsner says it is in "a long, scattered village now with more caravans than houses". Writing in 1968, Arthur Mee claims the tower was struck by lightning 700 years ago, and has not been repaired, though in 1976

Pevsner claims it has been "much renewed". Its pulpit is dated 1620.[2] Was it badly damaged by the flood and took over a decade to be replaced? It originally had a very rare pebble floor. The earliest parish register dates from about 1730, far too late to record the Great Flood. It includes a cure for a sick cow that involves dragon's blood. In the 19th century locals suffered much from the ague, a fever found in damp regions, probably malaria.

Mee claims "for years the north side of the churchyard was set apart for sailors drowned on the treacherous Brean Sands when Bristol's ships were constantly passing this way", so again echoing the treacherous south coast of Wales.

In the churchyard is a piece of heart-shaped iron, about 18 inches high, marking the burial of a sailor which bore the words: "The cruel winds and yawning waves hurried me to my doom, while wife and children dear waited for me at home." ... "The stump of the ruined tower was roofed over with gables, and in it still hang 3 bells that have been ringing for 400 years."[3]

Baker describes the flood damage beyond human life with:

"Added unto these peopled places, the loss of marshes, corn-fields, pastures, meadows, and so forth, more then can be numbered: the misery of it no man can express.

In this civill wars between the land and the sea, many men, women and children lost their lives: to save which, some climbed up to the tops of the houses, but the rage of the merciless tide grew so strong, that in many, yea most of the villages aforenamed, the foundations of the buildings being washed away, the whole frame fell down, and they dyed in the waters: Others got up into trees, but the trees had their roots unfastened by the self-same destroyer, that disjointed barnes and houses, and their last refuge was patiently to die.

"A lamentable spectacle was it, to behold whole herds of cattle, struggling for life, with the floods, Oxen in great numbers were carried away with the stream, and looked like so many whales in the sea: their bellowing made a noise in the water as if it had bin a tempest, and that ye Sea had roared. The flocks of sheep that are utterly destroyed by this land-wreck are innumerable, none knows the loss for the present but the owners of them: But the whole land will I fear feel the smart.

"A number of most strange shapes of danger did this monstrous birth of waters bring forth: of which [for the rareness] I will set down some, and none but those that are true."

Baker's account shows a mix of people who — despite the obvious surprise and initial confusion — tried to save themselves, loved ones and their goods. He also describes a man who was too confused by the disaster to think clearly, probably due to what we now recognise as shock, which must have been common:

"There was a poor man [a householder] dwelling in one of the villages aforenamed, having 7 children: who [in this general peril] not knowing howe to bestir himself, was desirous to save so much of his goods as possibly he could: But the violence of the stream multiplying more and more upon him: It came into his mind to provide rather for his children: his goods therefore he left to the mercy of that which hath no mercy, and loving one of his children above all the rest, his fear drove him to run about for the safety of that only. At last the danger that had round about [and within doors] set upon him and his family, was so great that he could neither defend that his dearest child, nor the rest, but having much ado to get life for himself, he left them and his whole household, some mile or 2 from the place where it was known to be kept, and so was preserved, for the cradle was not of wicker as ours are here, but of strong thick boards, closely jointed together, and that saved the infant's life." This echoes an accounts from Wales where a baby was likewise saved by floating in a tub.

In the fields "The ricks of peas in divers places being undermined at the bottoms, were lifted up mainly from the ground, and swum up and down in the whole bulk, amongst which a company of hogs, and pigs, being feeding upon one of the ricks, and perceiving it to go away more and more from them, they got up to the top, and there maintained their eating. Nay, which is more strange, conies [rabbits] in great numbers being driven out of their burrows by the tide, were seen to sit for safety on the backs of sheep, as they swam up and down and at last were drowned with them.

A poor shepherd likewise being in the field, [again, with no location given] some of his sheep were strayed from the rest, when the waters began to come in upon the Country which he perceiving, ran with all speed to fetch them in, hoping to save all; but before he had done, having much ado to save himself, he was fain to leave them, and with his bag and bolt, to climb up into a tree: there he saw the confusion of his whole flock: they swam to and fro bleating for help. He sat tearing his hair and beating his breasts: crying mainly out but could not save them: when they were all slain before his eyes, he wept then more bitterly to think upon his own tragedy which he saw was now to be acted: he feared drowning, yet he feared starving more than drowning: he had some victuals with him in the tree; but he knew not how long this siege of waters would keep him in that rotten bulwark. At length [when he was almost pinched to Death with cold] he espied a boat which the Country had sent out to save others, to that he called and in that recovered life."

Bristow

Baker's guide next directs us to Bristol, "and there behold as much cause of lamentation as in any place of this realm, that hath tasted of the like misery". But he claims it was "much about the very day" which is worryingly vague, when "did an arm of the North seas break in [at a spring tide] which overflowed not only the banks, but almost all the whole Country round about".

This clearly doesn't refer to what we call the North Sea but an area to the north of the city, so it must have been due to a breach in the South Gloucestershire sea walls. It is interesting that the Avon, with its famously high tidal range seems not to have been blamed. The modern river has been straightened at several places, especially downstream from the Hotwells where the spoil was hauled up to the Downs and used to fill in the quarries that were mined to provide stone for the expanding city in the 18th & 19th centuries. There was also an island called Dumballs which had silted up to join with the bank which was often noted as being an obstruction to shipping.

Latimer's 'Annals of Bristol in the 17th Century' provides the following account:

"A phenomenal flood tide occurred in the Severn on the morning of January 20th, 1607", again citing the modern dating system, "whereby the low-lying lands on each bank of the river from Gloucester downwards were inundated over some hundreds of square miles. The loss of life was estimated at 500, and a greater number of people were saved only by climbing upon trees, haystacks and roofs of houses. In Bristol the tide, being partially dammed back by the bridge, flowed over Redcliff, St Thomas and Temple Streets to a depth of several feet. St Stephens Church and the quays were deeply flooded, and the loss of goods in cellars and warehouses was enormous."[4]

Continuing south, Baker continues: "All *Brent-Marsh* is covered over : between *Barstaple* and *Bristow* the sea swelled up as high [far inland] as *Bridgwater*."

Bristol's previously cited Reverend Samuel Seyer helpfully suggests *why* Bridgwater was so badly flooded. In 1567, i.e. 40 years earlier, "a new cut was made in the river of Bridgewater ... by reason of a great compass or fetch about of the water of the said river. The sea banks or walls of the said river upon the north-east part thereof right to a tenement in the tenure of R- Popham, were so decayed and worn (notwithstanding yearly reparations done, to no small charges) that if the sea should have broken over, whereof the inhabitants of the country there, nigh to the same, were in great fear, it would have drowned about 10,000 acres of ground, beside other great harms which might have ensued thereof."[5]

So, again, the record shows that part of the problem was due to the failure to maintain flood defences, largely the result of the loss of engineering skills and the ability to organise large projects when the monasteries were shut. "It was therefore preventable and foreseen by the Commissioners of Sewars, Sir Hugh Pawlet, Sir George Speake, Sir Morreys Berkeley, knights, Mr Humphrey Coles, Mr Henry Portman, esquires, and others of the said country, that a new cut should be made straight over." It seems they spent a large sum straightening the river almost exactly 40 years before the Great Flood struck, allowing it to arrive faster and cause more damage.

Like Bristol, *Bridgwater* is an inland port, but the mouth of the River Parrett is wider and has more inlets than the Avon, and lacks the huge tidal surges which drove small ships into Bristol.

Barnstaple's flood damage also needs to be seen in context. In 1598 the town began to build a merchants' walk on the quay to accommodate the expansion of the town's trade. Works seem to have extended into the River Taw, deepening and narrowing the channel which annoyed the Earl of Bath. A court declared the work was a nuisance to vessels, and interfered with horses crossing the channel.[6] The details have been lost, but it seems likely that as it was not complete when the flood arrived, it aggravated its effects there.

Baker continues his account on Somerset with "All the low grounds are not only hidden in this strange deluge, but in danger [by the opinion of men] to be utterly lost. Whole houses were removed from the ground where they stood, and float up and down like ships [half sunk]." This again suggests that more people were lost, as their settlements were not counted, so can no longer be located. Because of the frequent flooding of Somerset, it is the most fertile region in these islands, which is why people endured the floods to live there. Accounts of housing in South Gloucestershire before the Second World war described how many houses had an upper storey so that when the region flooded they would move their furniture and beasts upstairs to wait for the waters to run away. Back to the storm:

"Their corn-mows and hay-mows are carried away with the stream and can never be recovered. All their fat oxen that could not swim are drowned; with such a forcible assault did the Waters set upon the inhabitants, what they who were in their houses, and thought themselves safest, could hardly make way for their own lives: by which means a number both of men, women, and children perished : their dead bodies float hourly above-water, and are continually taken up: It cannot yet be known, howe many have fell in this Tempest of God's fearful judgement."

Baker's account then returns to Bristol, which was full of traders from London, Wales and Ireland for the annual fair. "Most of the goods both of Citizens here in London that were sent thither, and of the inhabitants dwelling there, as also the Rugs & such other commodity which came from Ireland, to the faire of Saint *Paul*, which was now to be kept there, are [to an infinite value, and to the danger of many a man's undoing] utterly spoiled and cast away. Goods in dry-fats (vats?), and whole packs of Wares are daily taken up, but past all recovery ever to be good again.

But this reference to Bristol Fair is confusing, as Nichols and Taylor record a common council meeting of 24 July 1730 when local traders complained of the "great inconvenience and grievance they were under" in regard to the 2 fairs held ... there ... and sought relief from the outsiders threatening their trade ... The winter fair, commonly called St Paul's fair... was to be held and kept in the usual and accustomed place in this city for 8 days on the 1st day of March yearly so pleaded for relief".[7] Infuriatingly, this source fails to state where it was, though Seyer also mentions it. Latimer cites a newspaper describing entertainments at Temple Fair in January 1743, which fits.

Back to Baker: "This deluge hath covered this part of the country by the space of 10 miles over in length, at least up towards *Bridgewater*. Many thousands of pounds cannot make good the loss which the Country only hath hereby received. God grant there ensue no second misery upon this, worse to our Kingdom, than this Plague of Waters." It is unclear what is meant here. How could the situation be worse? Unless the author was expecting the plague itself, which could have been especially virulent in the exhausted, cold, hungry survivors.

Barker then moves from specific locations to personal stories, but without the people" names or locations, they are problematic. He provides a detailed, but geographically vague, account of "a gentleman dwelling within 4 miles of the sea [betwixt Barstaple and Bristol] walking forth one morning to view his grounds, cast up his eyes toward the sea-coast, and on the sudden was struck into a strange amazement, for he beheld an extraordinary swallowing up of all the earth, that had wont to lye visible and level to his sight: he could scarce tell certainly whether he stood upon the ground which he was sure the day before was his own: hills and valleys, woods and meadows seemed all to be either removed, or to be buried in the sea: for the waters afar off stood to his judgement many yards above the earth: he took them at first for mountains and heaps of clouds, but fear being driven back [with a courage and desire in him to save himself from this imminent danger] home comes he with all speed that he could to his own dwelling: relates to his wife what he had seen, and the assured peril that was preparing to set upon them, and [withal] counsels her and his whole family to bestir themselves, and to get [with such provision and goods as they could easily convey away] higher up into the country to some one of his friends. All hands presently laid about them [as if enemies had been marching to besiege the town] to truss up what they could and be gone."

This is a stark contrast to the image one might have of a flood approaching. Here the gent had time to return home, and for the family to pack up their valuables before fleeing to safety.

And yet "how swift is mischief when God drives it before him to the punishment of the world. All were labouring to bear away some of the goods but before their burdens could bee taken up, they were compelled to leave them, and to look about for their lives. The fardels which they had bound up to save from drowning, some of them were glad to leap upon to escape drowning themselves.

"The gentleman with his wife and children got up to the highest building of the house: there sat he and them upon 2 rafters, comforting one another in this misery, when their hearts within them were even dead to themselves from all comfort: they now cared not for their wealth, so they might but go away with their lives : and yet even that very desire of life put him in mind to preserve something, by which afterward they might live, and that was a box of writings, wherein were certain bonds, and all the evidence of his Lands: this box he got, with the hard adventure of much danger: he tied it with cords fast to a rafter, hoping what wreck whatsoever should overthrow the rest of his substance, his main estate should bee found safe, and come to shore in that haven.

"But as in the midst of this sorrowful gladness, the sea fell with such violence upon the house, that it bore away the whole building, rent it in the middle from top to bottom, they that could not get up to the highest rooms, were put to a double death, drowning and braining. In this flood the husband and wife lost one another: the children and parents were parted: the gentleman in this whirlwind of waves, being forced from his hold, got to a beam, sat upon that, and against his will rode post some 3 or 4 Miles, till at length encountering with the side of a hill [of which lighting place, he was joyful] there he crept up, and holding notwithstanding his safety still in his hand: there sat he environed with death, miserably pouring out tears to increase the waters, which were already too abundant: and to make him desperate in his sorrows, the tyrannous stream presented unto him the tragedy of his dear wife, and dearest children. She, they, and his servants were whorled to their deaths by the torrent before his face, & drowned doubly, in his tears, and in the waves. Yet because he should not be altogether the only slave of misfortune in this sea-fight, nor be more triumphed over then others that fell in the battle. At length a little to fetch life into him which was upon departing he spied his box of writings [bound as they were to the rafter] come floating towards him: that he ventured once again to save, and did so and in the end most miraculously came off likewise, with his own life."

It is hard to make sense of this. There is a sense of urgency, yet there was also time for the man to run back to his house, to raise the alarm and for his family to start packing. The flood was unprecedented, so they were unlikely to be thinking clearly, but it is infuriatingly too vague to reveal its location and hence to trace, especially so long after the event. Like several other stories which have survived, and given Baker's strong evangelicalism, it seems to be presented as an example of the rich suffering with the poor, and of greed being a factor in the loss of his family, making him appear as a modern Midas. But this was to ignore the reality of the time. Property ownership was a major source of income, and a safe long-term investment, so the loss of the gent's "papers" posed a threat to the family survival, on top of whatever harm was done by the flood itself.

An account is provided of another gentleman in the same county, this one newly married, who set out on a journey by horse to a nearby town "to be merry". His horse was saddled and he had put on one of his riding boots when "the wind came about, the point of his compass was changed, his voyage by land was to be made by water, or else not at all".

This is the only mention of the wind changing direction.

"For the sea had so surrounded the house, broken in, lifted of the doors from their hinges, ran up into all the chambers, and with so dreadful a noise took possession of every room, that he yet was all this while but half a horseman, trusted more to his own legs than to the swiftness of his gelding".

"Up therefore he mounts to the top of all the house the waters pursued him thither, which he perceiving, got astride over the ridge and there resolved to save his life, but *Neptune* belike purposing to try how well he could ride, cut off the main building by the middle leaving the upper part swimming like a Flemish Hoy [cargo ship] in foul weather".

"The gentleman being driven to go what pace that would carry him which he sat upon, held fast by the Tiles, and such things as he could best lay hold on, and in this foul weather, came he at length [neither on Horseback, nor on foot, nor in a vessel fit for the water] to the very town, where in the morning he meant to take up his inn."

This differs from the previous account, as the water here seems to have arrived with incredible speed, or perhaps this reflects the difference made by the previous man being outside, so he could see the landscape and the approach of the flood.

Strangely, this seems to be the only mention of noise. The flooding must have been incredibly loud: the roaring of the waters, the breaking of walls, trees, houses, the crashing of objects against each other, the cries of surprise, fear and pain from humans and animals.

Then the editor/publisher shows his hand with "A number of these strange tragi-comical scenes have been acted upon this large Stage of waters: It would swell into a massive volume to chronicle them all: let these therefore which I have delivered unto you, be sufficient, as a taste of God's Judgements: these are enough to make you know he is angry, let them likewise be enough to make us study how to allay his anger."

Baker ends his account of the West Country flood with: "Add unto these the overflowings in Herefordshire [not mentioned elsewhere, but suggests the flood reached further inland than other records state] Gloucester Shire, and in divers shires in Wales, bordering upon the sea, where many lives have been lost, both of man and beast: of all which when the particulars are truly known, they shall bee truly published to our country: till then make use of these."

Baker then confusingly jumps to the East Coast, "for just the same month of the yeare, week of the month, and almost day of the same weeke" he describes floods in March-land near King's Lynn. suggesting it was flooded at the same time, not a year later "though not altogether so violent and mortal as those in Summersetshire", which is highly unlikely, and no trace can be found of this in any other records. It provides some lively details of the flood and how people responded to it. But as these do not refer to the storm of 1606/7, the account can be found in the Appendix at the end of the book.

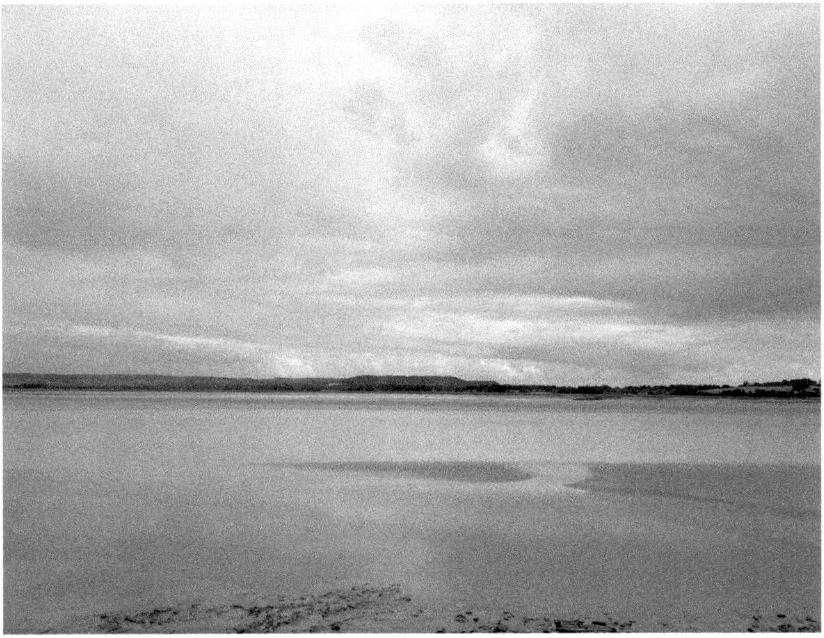

# CHAPTER 7
# OTHER SOURCES

The National Library of Wales provided the following article from its records, titled 'Notes on the Great Flood of January 20th", by C.H. James, citing the modern date of 1607. The article was to draw to members' attention to "a rare tract in Cardiff Reference Library", which has already been dealt with in chapter 4, 'Lamentable News...' It claims "The part of the Tract that may prove most interesting ... is that describing the actual results of the flood. The greater part of the Tract is a homily on the wickedness of the times and its result."

The language in the following has been modernised except for several town names.

This is followed by "In the month of January last past upon a Tuesday, the sea being very tempestuously moved by the winds, overflowed his ordinary Banks and did drown 26 parishes adjoining on the coast side, in the aforesaid County of Monmouth-shire ... all spoiled by the grievous and lamentable fury of the waters".

This is the only the second source that mentions winds. It then lists the parishes that were drowned, in no particular order:

Matherne, Magor, Wiston, Peterston, Portescuet, Redwicke, Lanwerne, Lambeth, Caldicot, Gouldenlife, Christchurch, Saint Mellins, Undye, Nashe, Milton, Romney, Roggiet, Saint Piere,

Bashalecke, Marshfield, Llanihangell, Lanckstone, Saint Brides, Wilfricke and Iston.

Allowing for variations in spelling, they are all on or near the South Wales coast, though Lambeth exists as street names in Cardiff and Penarth. Milton cannot be located in the region. Is this an error, or has it been lost or absorbed as settlements expanded in the ensuing centuries?

The list is similar to the place name that surround a framed print of unknown provenance or date behind the pulpit in the church of Nash. This version is curious as the letter 'E' is replaced by '£' sign. It includes Milton, Lambeth, Newtone, Brent, Bristowe, Barstaple and Romney. Bristol and Barnstaple were famous and the former in particular well recorded flood victims, but what of Newton and Milton? The former is a small settlement on the Wentloog Levels west of Peterstone and Milton seems to survive inland from Weston-super-Mare, so both would have been flooded.

But Lambeth, Brent and Romney seem to be in London and the South East, the former on the River Thames, the latter being marshes that are still at risk of flooding. These were also other mentioned in Baker's reprint, which suggests either one copied the other or they shared a mid to late 19th century date and source.

The 'Gentleman's Magazine' of 1762 cited a document in the Harleian Library which lists the affected places as Chepstow and the moor of Caldicot Cowbridge (not mentioned elsewhere), Goldclift, Matherne, Redclift, Newport, Cardiff, Swansey, Langhern and many other parts of Monmouthshire, Carmarthenshire and Cardiganshire. It estimates 500 men, women and children were lost, plus many thousands of farm animals. A milking maid mentioned in other accounts is placed in Merionethshire, and in Glamorganshire a blind man long bedridden had his cottage and bed swept into the open fields; just as he was about to sink he found the rafter of a house and was blown easterly to land. This direction seems odd as the tide was from the east.

Returning to the National Library tract, it states that the speed of the waters was greater than that of a greyhound, so about the same as that of the Severn Bore, and that it would take 5 or 6 years for the land to recover its fertility. Disaster was predicted: "there is no probability that that part of the country will ever be so inhabited again in our age as it was before the flood howsoever it hath heretofore bene reputed the richest and fruitfullest place in all that Country". But what country is being referred to here?

"Moreover the land overflowed with the Severn Sea is valued at above £40,000 by the year, only in the said County of Monmouth which is yet under the waters and be recovered again from them at the Lords good pleasure."

Again, without precise dates, this is hard to comprehend, but it confirms that the floods entered over the tops of the sea walls, but were unable to retreat via the same route.

There are also new anecdotes included such as: "these things are related as certain truths. As that a certain man and a woman having taken a tree fore their succour and espying nothing but death before their eyes : at last among other things which were carried along in the stream, perceived a certain Tub of great largeness to come nearer and nearer unto them until it rested upon that Tree wherein they were. Into which (as sent unto them by Gods providence) committing themselves they were carried safe until they were caste up upon the dry shore."

A "gentlewoman of good sort, Mistress Van" is named but not her location. "Before she could get up into the higher rooms at her house having marked the approach of the waters, to have been destroyed by them and destroyed, however her house being distant above 4 miles... from the sea."

The account then moves west to Glamorgan, with Mistress Matthews of Llandaff, upriver from Cardiff, who is said to have lost 400 English ewes. Many hundreds of men are reported to have drowned in this county, but many were saved by the actions of Lord Herbert and Sir Walter Montague, who sent out rescue boats for 10 miles. Lord Herbert was claimed to have delivered food and other necessities to those in need.

The document then cites Camden who described Monmouthshire in 1610: "Beneath this, lies spread for many miles together a "mersh", they call it the Moore, which when I lately revised this works, suffered lamentable loss: For when the Severn Sea at springtide in the change of the moon was being driven back for 3 days together with a South West wind and what with a very strong pirrie[?] from the sea troubling it, swelled and raged so high, that with surging bellows it came rolling and in rushing amaine [to the mainland?] upon this tract lying so low as also upon the like flats in Somerset shire over against it that it overflowed all, subverted houses and drowned a number of beasts and some people with all." This suggests the event was far less dramatic than other records provide. But this is confusing as the south west wind would raise the height of the tide rather than drive it back.

The author cites the marker in the plaster of the porch of St Bride's church Wentloog which dates the flood as January 1606 but mentions that a brass plaque in Goldcliffe church is dated 1609. He attributes the discrepancy to the change from the Old-Style dating system to the new, present one. But both these dates were from the old, Julian Calendar which had its year end in March, so by the New Gregorian calendar, this is the following year. The discrepancy between the 2 parishes seems to be that St Brides records the flood date, whereas Goldcliff seems to refer to the date of the plaque being erected. It must have taken time for the parish to recover and to commission, make and install the brass plaque. In the immediate aftermath of the disaster, the priority in the middle of winter would have been to provide food, clothing and shelter for survivors, to bury their dead and to carry out essential rebuilding and repairs before an expensive plaque could be commissioned and installed. This suggests the 1609 date was when the region could see the light at the end of the tunnel and was able to thank God for their survival.

James then claims the flood affected not only Glamorgan and Monmouthshire, but "the whole of the low-lying parts of England... viz Norfolk, Lincoln, Huntingdon and Kent", which seems to be from E.E. Baker, who published over 2 centuries after the disaster, and cites no sources, so this needs to be treated with care.

James then cites 'The Contribution toward a History of Swansey' by Lewis W Dillwyn, published in 1840, which in turn cites the 1607 pamphlet 'God's Warning to his People of England...' It was reprinted in 'Harleian Miscellany', stating the flood began about 9 in the morning as already cited in my chapter on this source.

This record claims Swansea was a victim of the flood, but Mr

Dillwyn noted: "I cannot find in either the Corporation or Parish Records that the town sustained any injury, or any other entry relating to this event than the following:

"Aputt? Swansea 17th day of February in the year of our Lord 1607." This seems to be a year and a month following other records of the flood, but as it refers to a plea for compensation, this makes sense. I have tidied up the original language to make it more readable, and the correct date is mentioned.

"Whereas the inhabitants of the town and borough of Avon hath sustained great losses and hindrances upon their sea walls by reason of the inundation of waters which happened the year 1606 whereupon the aforesaid inhabitance did make their petition unto the justices of peace at the general Sessions for some relief to repair the aforesaid walls the justices petition their case then Sir Thomas Mansell with others of the justices ... send their letter unto this town of Swanzey in the behalf of the aforesaid distressed people to request the good will and benevolence of all good and well disposed persons towards the relief of the aforesaid inhabitants of Avon to repair their sea walls.

"John David Edwards portreeve [the highest local officer, or warden) of the town of Swansea ... gave 20 shillings as a free gift & benevolence to the distressed inhabitants of Avon ... & Given into the hands of Hopkin Gryffyth one of the burgesses."

This account ends by noting that the Great Flood destroyed the old St Mary's Church, then standing near the site of the present Royal Hotel."

This suggests the damage to Swansea was not extensive. Like Cardiff, the main damage seems to have been to finish off an already ruinous church. This in turn seems to challenge claims that the flood caused serious damage further west where the estuary was wider, so the waters were shallower.

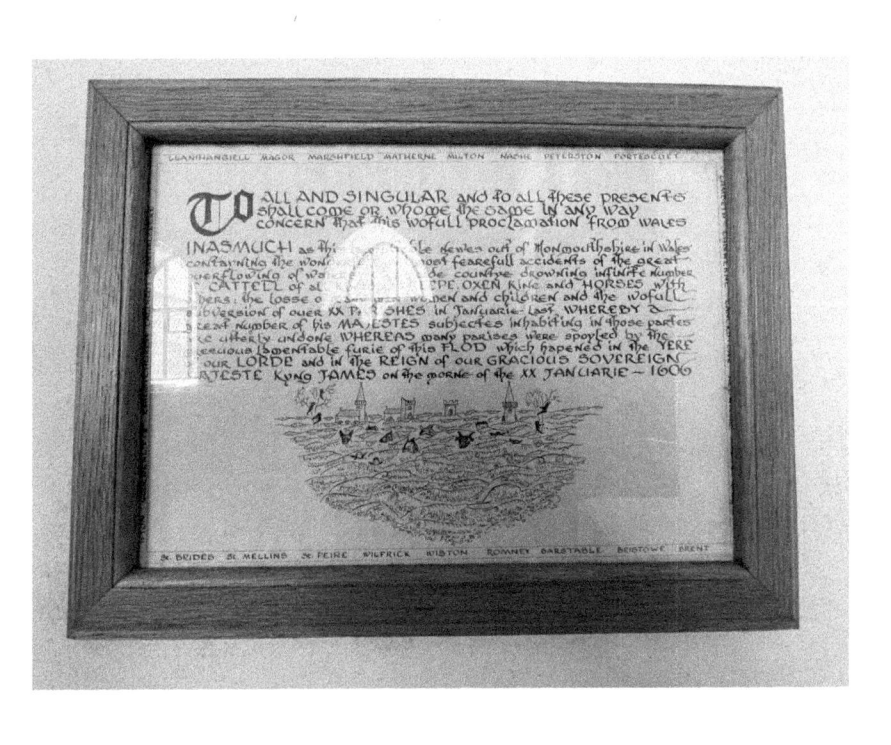

LLANFIHANGELL  MAGOR  MARSHFIELD  MATHERNE  MILTON  NASHE  PETERSTON  PORTESCUET

**TO** ALL AND SINGULAR and to all these presents shall come or whome the same in any way concern that this wofull proclamation from Wales

INASMUCH as thi[s] ...ble fewer out of Monmouthshiere in Wales containing the wond... ...ost fearefull accidents of the great overflowing of wa... ...be countrye drowning infinite number of CATTELL of al... ...EPE, OXEN KINE and HORSES with ...bers; the losse o... ...men and children and the wofull subversion of over XX P... SHES in Januarie last, WHEREBY a great number of his MAJESTES subjectes inhabiting in those partes be utterly undone WHEREAS many parishes were spoyled by the ...cious lamentable furie of this FLOD which hapened in the YERE OUR LORDE and in the REIGN of our GRACIOUS SOVEREIGN ...JESTE KYNG JAMES on the morne of the XX JANUARIE — 1606

ST BRIDES  ST MELLINS  ST PEIRE  WILFRICK  WISTON  ROMNEY  BARSTABLE  BERROWE  BRENT

## CHAPTER 8
# THE GREAT FLOOD TRACKED TO CARDIFF

In 2011 the BBC broadcasted a 'Timewatch' programme, 'Hidden Histories' in which Professor Simon Haslett of the University of Wales and Dr Ted Bryant of the University of Wollongong in Australia suggested the Great Flood may have been a tsunami, a phenomenon brought to widespread public awareness by the 2004 disaster in the Indian Ocean. They based this claim largely on the extreme forces unleashed by the event, especially the geological evidence of large rocks being displaced and huge areas of sand shifted. The site of the future Second Severn Crossing was washed away and the reservoir for Oldbury Power Station was carved from the cliff.

They suggested this was caused by an underwater earthquake in the Irish Sea, but this theory has been largely discounted, mainly due to the absence of any supporting evidence further afield. The 1755 Lisbon earthquake has also been cited; it triggered a tsunami which was felt in North Africa, London and Finland, and possibly as distant as the West Indies and Brazil. Cornwall was hit by a 3-metre tsunami, and part of Galway's city wall collapsed.

Haslett and Bryant found huge tidelines in Devon and Pembrokeshire, with the heights increasing as the channel narrowed, reaching 16 feet/5 metres in Glamorgan, 18 feet/5.5 metres in Somerset and 25 feet/7.5 metres in Monmouth, so following the pattern for the

Severn Bore. It penetrated as far as North Devon and South Wales to about 1.5 miles/2.5km, and about 15 miles inland to Glastonbury Tor. Once the wave collapsed, the water followed gravity to the sea, though on low regions such as The Levels, slow water flow meant that silt was deposited near the coast, causing river mouths to be higher than inland so the waters flowed backwards and deposited more silt.

Scattered through the references are records blaming the 1606/7 event on weakened or poorly maintained sea defences which had been designed and maintained by local monastic houses, but were neglected post-Reformation by private owners. Details of fatalities are scarce, as records were kept in parish churches, but priests were likely too busy helping parishioners, and many parish records have been lost to the flood, decay, Civil War damage and neglect in the following centuries. I was told by a historian that those who drowned would have been entered into church registers, but the flood swept away bodies, so many would never be located. The Severn was home to carnivorous pike, which must have been rare examples of creatures that thrived in the aftermath of the disaster. It was in the middle of winter, and people who lived far from help would have had their bedding, clothes and shoes soaked, so could have died of exposure in the following days and weeks. Without dry firewood, they could not keep warm or cook any food that they had.

Due to so many sites being affected, the following will begin in west Wales where the earliest records of the Great Storm survive, then track it up the River Severn, then to Bristol, which has the most detailed records of the event, and into deepest Somerset.

The timing was fortunate, at 9am, so people were already up and about, not torn from their sleep or drowned in their beds. But it was mid-winter, so victims likely suffered hypothermia from lack of dry clothes and dry wood for fires. Lords of the manors tended to live on higher ground, which may have delayed their initial rescue efforts, but this also meant they had the resources to provide food and shelter for victims.

The speed of the flood's arrival is shown by the account from Llanwerne, a village 4 miles from the shore of the Newport Levels, to the north of Goldcliffe, where a woman drowned as she could not climb to the upper floor of her house fast enough. Many people lived in small, single-storey cottages. But she was described as having £40 p.a., a significant income, which shows that wealth was no protection.

Churches were generally the largest and most secure buildings. Some in the path of the storm were above the flood line, so would have provided a refuge, a rallying point and space to to pray for survival. But in poor, isolated areas, churches were similar to the homes of the poor they served, so would have shared their parishioners' fate as the waters rose. In areas such as forests and the Somerset Levels, too poor to fund priests and maintain a vicar, some people were not part of a parish, living in small clusters of houses so they left no written records.

The following is the likely path of the flood.

The Church of St Ishmael is on a hill to the west above the Towey Estuary. It has been a place of worship for over 1,000 years and is now in the care of the Churches Conservation Trust. Though some distance from the present coast, it is surrounded by water and its cemetery to the south is close to what appears to be a silted former harbour. It was a chapel of ease for St David, Pembrokeshire with a 13$^{th}$-century nave, 14$^{th}$-century transept and 15$^{th}$-century north aisle. Its tower is a later addition, and seems to have been lowered at some point following storm damage. Could this have been due to the Great Flood? It is a mere 40 metres from the high water mark, and 400 metres from what seems to be the original settlement which supported the church. Several excavations have been made but no definitive confirmation has been made of its age. Suggestions have been made that it was lost due to the 1606/7 storm, but the parish website claims this was of 1607/8. Have they got the dates wrong or is this another storm, the records of which have been lost elsewhere?

This is from Rhys: "Four rivers have their aber/estuary in the inlet of Carmarthen Bay, watched by St Ishmael's Church on one side and Llanstephan Castle on the other – the two Gwendreaeths, the Towy and the Taf. Of these the Towy is a great stream; ... there is none in Wales, wild or sober, fairer in variety, richer in memory. The other rivers are tributary to it. To quote Leland, "at the low water marke a man may perceive how it ha(steth) to the sea on the sand is hard by Towe."[1]

The inlet is shaped like a starfish with 5 limbs, though 2 have sanded up. From Rhys again "Looking from Ferryside or from St Ishmaels at low water, you gaze upon miles on miles of sand. Then at times with a strong south-west wind behind, the tide comes up at a gallop and fairly races along the railway embankment. At spring-tides you meet it unexpectedly in the shape of frothy waves under the sandy railway arch, where the sand has been dry and the bits of seaweed have been brittle for weeks before; and I have seen it flow on and turn the high-road within and below the railway bank into a river. If you have been tempted out at low-water a mile or 2 on the beach where the cockle-pickers go, you may have to use strategy and skirt shoal-water to get safe home. In good aspects of the sun these tidal sand-deserts have a quite beautiful light bright colouring, with a dancing kind of radiance that suggests a mirage and some lost town like Hawton quivering in mid-air with more roofs and pinnacles than ever it had of old.

"To the inbound mariner these sands, beloved of small boys, cockles and flat-fish are a bother and a danger. If his schooner escape Cefn Sidan it may get sand-bogged at Warle Point. Hear the *Complete Sailor's* warning: "No vessel should try to make Laugharne or Carmarthen without a pilot." ... and of the Towey "To take a vessel 4 or 5 miles through an invisible channel not more than 3 cables wide, with a turning tide running 4 or 5 knots, and breaking sea on either side, can only be done by an old hand.""[2]

The Towey Estuary is described as having "broad shiny sands",[3] so should have suffered from the Great Flood, but no records can be found. Carmarthen is a fortified town on the north of the estuary, surrounded by wooded gorges. Thus, as with Chepstow in Monmouthshire, people could probably flee to higher ground once the alarm was raised.

Laugharne is an ancient borough and an isolated English-speaking area, where Flemings settled under the Normans in the early 12th century after floods drove them from their homes. It is near the mouth of the navigable River Taff which enters the estuary from the west, and was surrounded by fine farmland with a bustling market and monthly fair. Kidwelly is on the River Gwendreaeth which enters the bay from the east, and by the late 13th century it was surrounded by solid walls, and a century later was a major centre for trade and commerce in South Wales. It was surrounded by wide meadows and marshland behind dunes, so must have flooded frequently, providing welcome fertility to the soil for grazing before silting became a problem in the 18th century.

These towns' positions inland suggest that they, like Bristol, were protected from all but the most extreme weather and from raiders.There seems to be no record of the area being struck by the Great Flood, but it suffered outbreaks of plague in 1604 and 1606, so it seems the storm would have struck in the wake of these disasters, leaving fewer to be victims or to leave records.

South east of the Towey Estuary is another region centred on Burry Port that has long struggled to survive, as Rhys describes: "The constantly growing and multiplying sands of this estuary make a Sahara of the coast-line between Pembrey and Llanelly. It continues about 6 miles beyond the "Nose" over Pembrey Burrows and Towyn Burrows to Towyn Point. Beyond the end of it, at low tide, lie Cefn Sidan sands, 2 or 3 miles more. Here they say once stood a fair city. Traces of foundations of walls and stubborn roots of trees are still to be seen at low tides, after a heavy freshet in the Towy. How far the lost city many be traced to notions of the washed-away village of Hawton, which is shown in Saxton's and Speed's 17th century maps of the county, it would be hard to tell. Hawton lay, however, on the other side of the Gwendraeth, under St Ishmael's".[4]

Rhys continues: "Coasting vessels long had a very warrantable dread of the treacherous sand-bay into which flow the Gwendraaeth, Towy and Taff Rivers. It is hard to get even a small yacht through the sands up to Kidwelly for you have a snaky track to negotiate that is only seen for what it really is at low water. The navigable way is bare half a cable in width, and on either side, at flood, there is about 3 feet of water over the sand. "Nothing bigger than a coal barge", says "The Complete Sailor"."

Rhys, writing in 1911, described the shifting geography of the region with: "The sea has won a long tide's reach on the land almost within living memory. A print ... at Swansea Library shows ... the vanished old parsonage house, Oxwich, washed away by the sea, about 1805. It stood between the church and the tide-line, which has now crept up to the church-yard."[5]

Walesonline.co.uk of 10 Oct 2019 claimed a Swansea village was "washed away by a tsunami" and was now buried just off Mumbles Head. Modern maps show that the west of Swansea Bay peters out through a nature reserve — often indicating uninhabitable land — with several rocks extending into the Bristol Channel, suggesting land used to extend further there than now. Records name it as Green Grounds and that a bridle path led from Penrice Castle, north of Oxwich, to Margam Abbey past Mumbles Headland but had been swept away by a storm. But this seems to understate its size.

On the Cassini Historical Ordnance Survey Map of 1830/1, Swansea Bay was very different from today, with about half the area showing drains leading to the Inner Green Grounds, with the Outer Green Grounds marked slightly inside the mouth, approximately level with Mumbles Head, so the land area was much higher than at present. It must have been subject to frequent flooding, so was marsh rather than arable farmland.

Green Grounds was cited as recently as the 1890s when Sir John Morris planned to build a quarry on Mumbles Hill. But was stopped when the Misses Angels, a pair of sisters, claimed ownership of the land from All Saints Church to the Headland. They owned a map showing their land including a large meadow, extended into the bay. A survey of East Gower also showed a rock allegedly from their family's old house.

At the time of the Great Flood, it seems many churches were in poor condition, the result of post-Reformation neglect and subsequent underfunding, which put them at higher risk of storm damage or loss in the 16th and 17th centuries. Penmaen on the Gower had a Church of St John the Baptist, but it seems to have been moved in the 13th century, a period of extremely stormy weather, to provide a medieval hunting park, Parc le Breos. There is a church in the burrows nearby which was abandoned as it was "be-sanded" in the early 14th century, as the extreme weather continued.[6]

Some local churches have provided records of the Great Flood. But St Mary's in Swansea's history is so poorly documented it is unclear whether the present building is the 5th or 6th on the site. Like many others in the region, the first church was probably built by the Normans and rebuilt in the stormy 14th century but was in such a poor state by 1739 that the nave collapsed. Its replacement is claimed to have been "small and cheap" and failed to cope with the surge in population oat the Industrial Revolution. Sir Arthur Bloomfield designed the reconstruction in the 1890s, but this was destroyed by German bombs in 1941and was not replaced till 1959.[7]

Records survive from Aberavon Beach near Port Talbot to the east of Swansea, where the sea walls were damaged by the 1606/7 flood. This prompted officials of "Swansey" to give 20 shillings towards their repair.[8] The document recording this was dated February 1607, probably the Old Style, suggesting it took over a year to settle the matter. Unfortunately for this account, the source makes no mention of any casualties.

Ernest Rhys provides an extensive, wonderfully detailed account from 1911 of the coastal region to the east which has struggled for centuries to contain the ingress of sand, so it is worth quoting at some length:

"It is a wild stretch of coast that runs westward from the estuary of Ogmore River. The battle between sea and land is fought there with endless change of fortune: the sea hurls up billows of sand to choke the fields and bury the houses; the land sends out deadly ridges of low rock to the murder of ships that pass in the Severn Sea.

"Newton Nottage here stands back from the sea, with a broad belt of half-clothed dunes between. The village is a survival of other days, worth making acquaintance with, whose people are people of character and humour, and of a canny quality too … The church … is original … and its tower is not like other towers, but has an "air" not easy to realise as it stares across at the Merthyr Mawr warren like some amazed and amazing creature spawned by a primitive world."[9]

"Westward from the Black Rocks consists of drift sand and rolled pebbles. This flat beach is divided at Newton Point and Middle Point by Skers or projecting ridges of low rock. Each of these spits, as well as the higher point at Porthcawl … is probably continued into the Channel to the south and east under the names of the Patches and the Tusker.

The latter rock has a beacon on it, and is especially dangerous from the deadly skers which open out at its western end; upon them the tide sets with a heavy break in rough weather. When the foundations of an old house in the neighbourhood were re-dug, a drift of sand like a raised beach was found below them. The sea had been there before; just as certainly as the sea is gaining upon this coast again.

"Dutch vessels northward of their true course, and deceived by the similarity of the soundings, sometimes came up the Bristol instead of the English Channel — an error often fatal before lighthouse days. A flat stone in Newton churchyard commemorates the loss of a young family, 3 sons of J.S. Jakerd, sent from Surinam in the planter's own ship westward bound to Amsterdam, wrecked on the night of the 3rd June, 1770. Many soldiers lost in one of the transports for [from?] Bristol in the Irish Rebellion, 1798, were buried in Cae Newydd, Porthcawl. In a touching act of kindness towards strangers, "the plough for long spared the turf above them".[10]

Rhys continues: "In the sandy desolation east of Porthcawl lies all that is left of Kenfig town today: a scattered hamlet with the last fragment of a castle tower thrust up like the hand of a buried man through the sand. Entering Kenfig today by its loneliest road from the east one hardly believes in its existence as a live place, so dispersed and silent among its sands. Even its church tower is not at once visible, and the few houses are hidden in the perspective, one behind another ... Passing by the first old inn, you cross the sandy turf, find a sort of sandy lane or street, and make the circuit then of the churchyard...

"The old Town Hall is now the Prince of Wales Inn, and there may be seen the old town charters, and other documents, bearing witness to the Kenfig that was. Today it is a place apart, like no other I know; an amazing place with a heroic record: the struggle of the community during hundreds of years with an irresistible army of sand, whose tents and entrenchments you see all around it. "The sand came up like snow and buried the houses", the old people will tell you. Today the drift still goes on; while 600 years ago it was coming steadily inland, each storm bringing the sand higher. The first mention of these inroads is preserved in the record of reduced rent for a warren called "the Rabbits' Pasture," "because of the great part is drowned by the sea."

This was in the year 1316. More than 200 years later the traveller John Leland wrote: "There is a village on the East side of Kenfik and a Castel, both in Ruines and almost shokid and devowrid with the Sandes that the *Severn Sea* there castith up."[11]

"Yet it was (for that day) a large and important town that waged this losing battle, as we learn from the Kenfig ordinances drawn up in 1330 ... It was a walled town with paved ways; it had from 7-800 inhabitants. It had a river up which came merchandise and timber ... It had a stout paternal government of portreeve and burgesses. Most of the houses were of wood, and the town was not only drowned in sand, but was burned again and again; for the nest of fighting men up at the Castle were for ever waging war, and each battle that came up against it first burned the town by way of recreation. It was for ever being rebuilt, and later on more soundly: the ordinances certainly disclose a most respectable state of things".[12]

The town ordinances reflect those of other established towns, enforcing standard sizes of bread and the quality of foodstuffs, and each householder had to keep the street in front paved and clean. ... There were also laws on maintaining the land, with a ban on cutting or pulling up plants "or any other thing that may be to the ruin, destruction and overthrow of the said borough". But in the reign of Elizabeth, the problems increased as the poor citizens "doe yearly fall in arrearages and losses... by reason of the overthrow, blowing and choaking up of sand in drowning of our town and church", which has vanished, "with a number of acres of free lands, and the lands within the town" which people were still paying rents on.[13]

Rhys also describes the nearby Kenfig Pool which swallowed up another town where "teal and wild duck, as they take wing or skim the water, lead their own life, undisturbed by the ghosts of the past". It is a curious place, of fresh water below sea level surrounded by encroaching sands. In later years, Kenfig descended to become a haven of bandits, smugglers and wreckers.

The records of Margam Abbey claim that one of the first serious invasions by blown sand on the Glamorgan coast was in 1384.[14]

Llantwit Major is listed as being on the Severn Channel coast in the Vale of Glamorgan, but it is inland from a small pebbly beach, reached by a rocky valley about a mile long which allowed the export of lime in exchange for wine and hides. But the port was damaged by storms in 1584 and 1607. Yet again, was this Old or New Style? If the latter, it seems to have suffered from the Great Flood.

East of the above, the parish of Flemingston is about 5 miles south of Cowbridge. Evans claims only the font survives to confirm this. The 14th-century church has a stone effigy of Joan le Fleming; the text in Norman French grants 40 days' pardon for anyone who prayed for her soul. The church plate bears the date 1607.[15] But again, is this the old or the current dating system? If the latter, was this an offering for the unnamed donor surviving the Great Flood?

To the east, on the coast is Aberthaw where limestone was mined for Victorian industries, so it was a busy port. Its river was the same size as Rhymney to the east of Cardiff and its port was inland. It was bigger than Llantwit Major and Porthkerry; its merchants traded with Spain, Portugal, France, Bristol and Somerset, especially when the latter 2 were damaged by a storm surge of 1584 and by the Great Flood. But mid-channel dredging has reduced the extensive sand dunes there. The now-redundant power station is protected by the extensive sea defences similar to those of the Wentloog Levels.[16]

To the east, on the coast, is Rhoose Point, the most southerly point of Wales, so it must have been badly affected by the Great Storm.

Further east, on the coast, about 9 miles south west of Cardiff is the town of Barry. It is mostly a beach resort for Cardiff, and its history is hard to ascertain. Wikipedia provides some detail. It was a Stone Age settlement, Romans had successful farms there with a thriving port, but this was followed by Viking raiders and Norman settlers. It has a harbour and extensive docks, so it was heavily dependent on the sea. It was also, like many settlements in South Wales, a pilgrimage site. Also like much of South Wales, it is on a river which flowed through extensive marshland, so it must have suffered from many storms, especially that of 1606/7, but there seems to be no record of it or any damage caused. The mostly likely explanation for this silence is that the industrialisation and commercial development has erased all trace, and there is no major church to provide contemporary records.

But on YouTube is a piece by Graham Loveluck-Edwards, 'The

Romans in Barry', in which he helpfully mentions the flood. He says one of the best-kept secrets is the Roman remains on the Knapp at Barry, a promontory with a marine lake and gardens jutting into the Severn. He describes it as a natural harbour when Marcus Aurelius broke with the Roman Empire to found an independent state in Britain with a fleet on the South Wales coast. He says a vast bank of marbles was deposited there in 1607 which closed off the port into Cold Knapp. This was likely the year of the Great Flood.

Midway between Barry and Penarth is the small settlement of Sully, site of the smallest castle in Glamorgan, and the parish church of St John suggests it is a very ancient foundation. To the east near the coast is Sully Island which is only accessible at low tide via a rocky causeway. It sheltered St Mary's Well Bay where local traders tried to avoid paying import duties to the Crown. It must have been swamped by the 1606/7 flood.

The town of Penarth is now a short train journey from central Cardiff, and its parish church rises above the entrance to the docks. Its height should have meant it was dedicated to St Michael, but until 1543 it was owned by St Augustine's Abbey of Bristol, so it shares that dedication.

Evans describes it as "a very pleasant watering place, a seaport, and residential suburb of Cardiff... Standing on the estuary of the Ely river, ... built around a bold headland and commands splendid views of the Bristol Channel and the Somerset coast, as well as inland... it is hard to believe that its population in the middle of the last [19th] century ... was less than 100 people.

"Its extraordinary growth was due to the incapacity of the docks at Cardiff to deal with all the coal coming down from the valleys; but the port... of Penarth, with its dock area of 26 acres, has never interfered with the pleasant amenities of the town."[17]

The most significant family in the area was yet another branch of the Herberts who lived in a Tudor mansion on the hillside at Cogan Pill. They were said to be involved in piracy, but at the time the lines between this and of legal trade were often unclear, as shown by the successful career of Drake and others. When the storm struck, the hilltop must have provided shelter for any survivors. It was still a small rural fishing village until the mid-19th century. But Evans adds "It occupies the site of a former church which tradition says was Norman built and had become sadly decayed". This suggests few records were made at the time of the flood.

The impact on the city of Cardiff is poorly recorded, and it is now impossible to imagine due to the huge changes to the city since the event, especially the 19th-century building of the docks, fed by a canal diverted from the Taff at Black Weir, the north of the city centre. South of the city centre the rivers Taff and Ely — the latter with its modern marina — feed into Cardiff Bay which is now protected from the Severn's tidal surges by a barrage and Portway. The huge levels of shipping mean the bay was heavily polluted within living memory, and the 3 redundant docks have signs banning swimming. But locals tell me that decades ago they earned pocket money by catching crabs in the estuary and selling them. All this means it is impossible to imagine what the area looked like at the time of the Great Flood.

St Mary's Church in the centre of Cardiff is commemorated by a street name. Accounts claim it was ruinous at the time of the flood and had long been falling into the River Taff, with many coffins being washed away. It stood at the South Gate from 1107 to 1620, so it must have survived but been further undermined by the Great Flood. In 1620, her chapel of ease, the 12th-century St John the Baptist near the castle, became the town's main church, but St Mary's parishioners struggled on. It was declared a ruin in 1678, her tower collapsed soon after, and the last service was held in what was by then a roofless building.

But this story fails to explain why the site of the church is now several blocks from the River Taff. In 1850 IK Brunel diverted the river to build the Central Railway Station, creating Westgate Street at the same time. St Mary's former site became the Theatre Royal, which was renamed the Prince of Wales Theatre before becoming a pub, which bears the outline of the late church on its outside wall. An old man told me that the land reclamation mostly used rocks that had been ballast in ships returning to port, which seems to be a good way of disposing of them.

St John's is further inland and further from the River Taff than the mother church, but there is no record of it suffering in the Great Storm of 1606/7 or that of 1968. But this is curious, as a policeman told me of how his involvement in the rescue of a circus that was camped at Sophia Gardens on the other side of the river, not far from the church. This latter storm led to the construction of the extensive embankments which now defend the city against the river's misbehaviour. For now. Parks and playing fields on either side serve as flood plains if the barriers prove inadequate.

The pedestrian suspension bridge over the River Taff near Pontcanna Fields has a weir which raises the river level to allow water to be diverted into a canal which flowed beneath the city centre but has recently been opened up. It feeds the docks and then empties into the mouth the River Severn, and has been part of the city's flood defences since the 19th century.

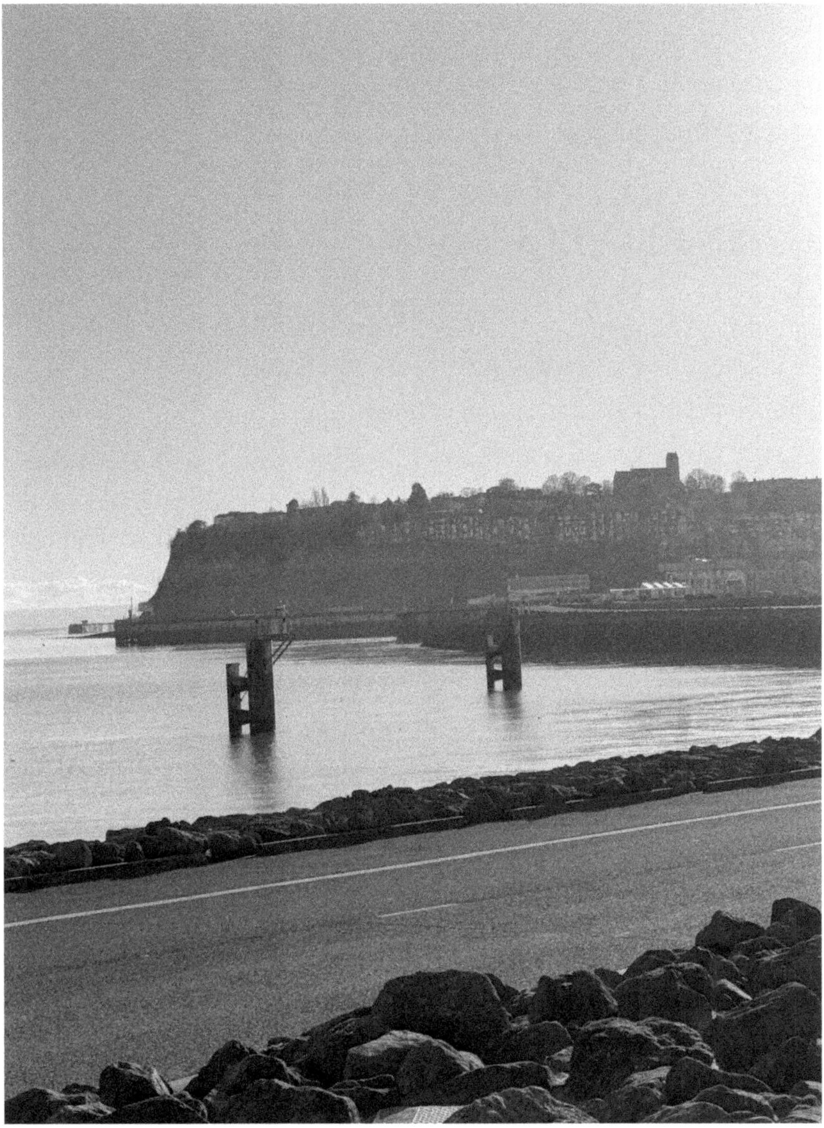

## CHAPTER 9
## THE FLOOD TRACKED TO GWENT LEVELS

I discovered the story of the Great Flood many years ago when I visited Redwick church and noticed the flood level recorded on the outside of its porch. I have searched in vain for others elsewhere, for it seems the South Wales Levels are the only places where such objects survive. The disaster should have been recorded in parish records. But clerics often jotted down details of births, deaths and marriages on scraps of paper or parchment, to be entered into the parish registers later. It seems that in the aftermath of the disaster, these practices were not followed. If they were recorded properly, the registers have since been subject to loss and/or damage from neglect, damp, closure of the churches, and in some areas, from subsequent floods.

This region is the best-known victim of the Great Flood, where several churches commemorate the disaster with plaques marking the height of the waters. It stretches from the east of Cardiff beyond Newport, curving north east along the river with the Second Severn Crossing flying above the coastal edgelands. Near Cardiff the region is mostly an industrial wasteland. A giant metal recycling centre looms like a prehistoric beast whilst the commuter traffic speeds past, catching glimpses of the mud flats and traveller camps on the riverside where their sturdy ponies graze and gaze in bemusement.

The Caldicot Levels are crossed by a varied mix of tidal ditches/drains called reens, the origin of which term is shared with Rhine and Rhonda, with their exits to the sea having the Saxon name of gouts. The biggest is Monks Ditch which was built by the clerics at Goldcliff Priory. It rises at Wentwood to 8 feet above road level before passing the tiny, now- ruined church of Whitson and on to the sea.[1]

To the east of Newport is the Wetlands centre, a haven for wildlife and humans. Chalet-style housing sits alongside Victorian villas and old farm buildings sinking into the landscape. The area is largely deserted on weekdays.

The number of impressive churches, many founded by Normans from Bristol, are evidence of the region's wealth in the days when the Severn provided huge quantities and varieties of fish, and sheep grazed on the salt marshes, which were flooded to attract waterfowl to overwinter and provide fresh meat. Roman embankments are evidence of busy maritime trade supporting their settlements at nearby Caerwent and Caerleon. All of this shows the region supported a surprisingly large population, involved in agriculture, fishing and local and international trade.

But all this came at a price; the sea was a generous neighbour, but sometimes a fickle, violent one. It provided food, shelter and employment, but as in 1606/7, it could turn hostile.

It seems that climate change is increasingly driving the Severn's behaviour to threaten lives and property, which is deterring investment and settlement. It is also a region largely neglected by historians apart from the author/historian Fred Hando, whose walking tours have been expanded and updated by Chris Barber. The area has 5 churches with markers recording the flood levels, which seem to be unique survivors. Part of the explanation may be found in the 'Buildings of England' series by Pevsner, which often record the interiors of churches as having been "scraped" by Victorian improvers. This raises the possibility that there were other markers which served as reminders of the great tragedy, as a commemoration of the dead and as a warning of God's power to punish his people. If so, were the walls scraped out of ignorance, or was the need to warn worshippers no longer relevant in the Industrial Age, when engineering and science seemed to promise that such tragedies had been consigned to history?

The following sites are listed from west to east.

The first Ordnance Survey map of Cardiff and Newport reveals significant changes in the region, especially along the coast. Penarth "Roads", possibly a reduction of "Whale Roads", i.e. navigable coasts, extend about 2 miles from the coast to Cardiff Grounds, with Peterstone Flats marked slightly further out and extending about 4 miles.

Between the mouths of the Taff and Rumney the coast is marked as "The Orchards", suggesting the region has lost coastal farmlands. An area on the coast is called "The Splott", claimed by locals to be an abbreviation of "God's Plot", land used to support the parish church. It is separated from Lower Splott by a sea bank, so this was prone to flooding, with a salt marsh to the east. Across the Rumney mouth beyond another sea wall is Little Wharf; beyond a reen, this becomes Peterstone Wharf, and on the west of Uskmouth is Sealand Wharf between which old sea walls run. South of the mouth of the Usk are dangerous areas for ships called The Patch, and The Split which is separated from an unnamed region by "Shord Passage", beyond which were the English Grounds, probably used by coastal fishermen. It seems this was a dangerous region for ships.

Cardiff was drastically changed by the construction of its port from the 19th century, but the present Cardiff Bay was previously moorland similar to Penarth Moors on the River Ely to the west. South of the main docks are flats and ledges, with Pengam Moors beside the meandering, tidal River Rumney River on the city's eastern edge. Maps show the remains of a wharf and disused tip which the Wales Coastal Path detours around. This area seems to have been a victim of the Great Flood.

The Church of St Mellons adjoins the area, slightly higher, inland to the north. Part of its boundary follows the River Rumney which meanders through playing fields and nature reserves, suggesting it is still a flood risk. St Mellons is on higher ground than most others and the parish is now part of the leafy Cardiff suburb of the same name, though the history of this name is unclear. Old St Mellons is to the north, across the river, and still mostly open land. It is north east of Rumney, seems to have been named as another victim of the storm. Only a mound survives of Wentloog castle. The region runs mostly between the railway line and the coast. The name Wentloog is suggestive of links with the Low Countries, so may have been another site where Flemings settled under the Normans, though 'loog' in modern Dutch means 'lied', so perhaps this source is a non starter. It is some distance from the coast but Faendre Reen runs through it, so there must have been another flood risk.

Back on the coast is the tiny settlement of Newton, protected by Rumney Great Wharf with its breakwater. It was listed as a victim of the flood. Wentloog Road leads to Broadstreet Common which runs alongside the long oblong stretch beyond the coast called Peterstone Great Wharf, now a salt marsh that may have been used for grazing the sheep that Charles 1 became so fond of. Near the end of the wharf is the Church of Peterstone/St Peter Wentloog. The first church here was built by Mabel Fitzrobert. It was given to the Priory of St Augustine, now Bristol Cathedral, but she is buried in St James' Priory in the city.

Sadly the church is now in private hands, with warnings of dogs on the loose. Barber claims there is an inscription stone on the north east corner with "THE GREAT FLOOD JANUARY 20, 1606" in lead, 5 feet 6 inches above ground.

Barber, writing in 1987,[2] mentions the repairs to the previous winter's damage to the sea walls, a reminder that life in the region still demands watchfulness and maintenance as the sea never sleeps. He writes "the tower, seen from any standpoint is noble ... in the pinnacled porch I found again carved heads of royal, angelic and holy aspect." The church seems huge for such an isolated site, but the parish formerly extended far across the mudflats, there was a hugely profitable fishing industry, and at the end of the Gout Fawr — Great Gout — was the New Quay where farm produce was shipped to Bristol.

Barber cites a local man who claimed the parish once extended beyond the present wall to another beyond the mudflats, which would help explain the huge church and its proximity to the modern water's edge. He also mentions a small bay with cut stones, all that remains of Peterstone Great Wharf where farm produce was shipped to Bristol.[3] This explains the stretch abutting the coast which is too clearly marked to be a natural feature. The trade with Bristol must have been immense to fund it, or it may be Roman remains where ships crossed the Severn from Portbury for supplies of Mendip lead.

The nearby Six Bells inn with its huge fireplace — 13 feet wide — is sadly in an advanced state of neglect, but shows there was a long-standing, sizeable and thriving community here.

Northwards and inland is the substantial settlement of Marshfield, between the railway line and the motorways. It is a $12^{th}$-century church in a large churchyard with many large evergreens and the village is now dominated by single-storey villas. The church was built by Mabel Fitzrobert, Countess of Gloucester and Glamorgan, the eldest of 4 sisters, the rest of whom became nuns and benefactors. It has a flood marker on the rear of the church.

About 3 km to the east, set back from the road is St Brides/Bridget, near the aptly named Sealand Rheen and inland from the West Usk lighthouse. St Bride's is claimed to be one of the most beautiful churches in Gwent, and its interior is one of the most fascinating. But its tower is leaning, and its parishioners ageing, so their numbers are falling. It is dedicated to an Irish virgin nun and abbess who is a patron saint of Ireland and Scotland, her cult widely spread by Irish missionaries, which helps explain this coastal foundation. But her wide range of patronages — except perhaps of boatmen — fails to explain her presence here, though her role in converting pagans to Christianity seems to fit well with that of the nearby St Thomas the Apostle. It is in the midst of a once-flourishing agricultural region. On the main wall of the porch is a plaque which was fortunately not covered up when the porch was replastered, commemorating the tragedy that was still seen as important to the parish.

It records "THE GREAT FLUD/ 20 JANUARY/IN THE MORNING/1606" with the churchwardens' initials. I was told this was higher than the Great Storm of 1703, as described in detail in the book by Daniel Defoe. But it only reached 4 feet rather than the 6 claimed for 1606/7. The churchwarden lamented the poor maintenance of the reens and described a reservoir towards the coast which should have been cleared. We discussed the likelihood that if the flood had been a storm, the heavy rain would have poured down the steep valleys to the coast. But this would have been mitigated by the fact that the coal mines did not exist at the time, when the slopes were still heavily forested, which would have delayed the runoff.

The Great Flood was followed by others in 1708 and 1800 which seem not to be recorded elsewhere.

The church pamphlet claims its last services were held in 1989, and in 1991, it was ordered to be demolished. But only 5 years later, it was restored using lottery funds. It was re-dedicated on 1 December 1996 and an extension added in 2002, but attendance across the region continues to decline, so given the high cost of the repairs needed, its future is again uncertain. Its design is compared with Peterstone, but any repairs have been piecemeal, reflecting its long-term decline. Surprisingly, the pulpit and pews came from St Mary, Swansea in 1899, suggesting the parish was too poor to buy them new.[4]

There is another church with the same, unusual, dedication at Netherwent, to the north of Magor, so it was also in the flood region. It is another curious sighting of the saint, but again suggests it dates from the early Christian era. It stands alone in a field near a village deserted after the plague.

Continuing east, and turning inland on the outskirts of Newport, the next destination of interest is the Tudor mansion Tredegar House, described as almost close enough to hear the sea. The park extends to Bassaleg, another site mentioned in records of the Flood, and it is sheltered by Gaer Hill to the north east, with the River Ebbw rushing to the sea "with pebbles in its throat". Closer to the mansion is an ornamental pond that after a tortuous route feeds the Sea Walls Reen which eventually joins the Tredegar Reen and the Severn. The mansion was owned by the Morgan family from Tudor times, possibly making it unique, given how many families have died out over time.[5]

At the mouth of the Usk, Robert de Chandos dedicated a Norman priory to the Virgin Mary and Mary Magdalene. The funds also allowed the founding of Christchurch, Goldcliff and Nash. The monks came from Bec in Normandy and established the salmon fishery and built the main drain, Monks Ditch. The priory was built between the farmhouse and the sea where hazelnuts have been found, which played a major role in human settlements as the land emerged from beneath the waters, as recorded by Gerald of Wales. They provided food and building material. When the monasteries were closed, Goldcliff was granted to Tewkesbury, and at the Reformation, to Eton College. In the 14th century, the region was known for wrecking and plundering ships, which continued into the modern age.

Across the River Usk to the east of Newport are the Newport Levels, a series of tidal ditches and reens which drain the moor, especially the Goldcliff Pill with the tiny church beside it. Robert Fitz-Martin of Compton Martin gave half a hide of land in his parish and the same amount on Mendip to the Abbey of Goldcliff.[6] The original priory stood on the coast, 60 feet above high tide where Hill Farm is now partly built from the remains. The Ordnance Survey map shows a stretch of shingle extending from this point into the channel, and another area in the river to the east, again suggesting the coast was further into the river. Barber claims that prior to this, the Romans led by Centurion Sartorious built part of the sea wall there, and left a stone which was found in 1878.[7] He also claims that the unusual name is derived from the cliff being made of a red sandstone base with limestone above which glitters in the sun. Pevsner claims the 2nd Legion land extended as far as Goldcliffe 3 miles away.[8]

Nash church is called the : "Cathedral of the Moors", though the title is also claimed by Magor and Peterstone. It belonged to Goldcliffe Priory from 1349. It seems it was much larger, with a north aisle; during the Second World War, congregations were up to 400.[9] There is a marker on the tower buttress recording the height of the 1606/7 flood but giving the year as 1607, potentially causing confusion. In the early 20th century a terrible storm drove a ship onto the rocks at Nash lighthouse. They found the crew of the Bristol trader lashed to the mast, dead. The bellringer claims he was forced to announce the service standing among the victims.[10]

St Mary Magdalene, Goldcliffe is a small church to the east of Nash, built 1 mile inland on the Goldcliff Pill to the north west following the part-destruction of the priory by a flood. Its limestone construction may mean it was made from the priory buildings. The church is approached by a fine avenue of pollarded lime trees, but it is poorly signposted, and there is no information on services or parish contacts, so much has its congregation declined. But on its north wall, 2 feet, 3 inches above the chancel floor is the most impressive church marker, in brass, dating from 1609, suggesting it took some time for them to recover enough to commemorate the tragedy.

The text is:

1606 ON THE XX DAY OF IANUARY EVEN AS IT CAME TO
PAS IT PLEASED GOD THE FLUD DID FLOW TO THE
EDGE OF THIS SAME BRASS AND IN THIS PARISH
THEARE WAS LOST 5000 AND ODD POWNDS BESIDES
XXII PEOPLE WAS IN THE PARISH DROWND
GOLDCLIF- JOHN WILKINS OF PIL REW AND
WILLIAM TAP CHURCHWARDENS 1609

The flood of 1606/7 was described as "caused by a serious transgression of the Bristol Channel and it affected coastal flats of Monmouth, Glamorgan and part of Cardiff".[11] Parishioners also presented a plate, paten and chalice, with their names and the date of 1609, a matter other sources fail to mention.

When Deerhurst near Tewkesbury and other "alien priories" were confiscated by the Crown and given to Eton College in 1442, Edward VI restored the priory, eventually giving it to Tewkesbury on condition that the abbey maintained 4 monks at Deerhurst. He compensated Eton with the gift of the "alien" priory of Goldcliffe, which had been given to Tewkesbury.[12]

A short distance to the east is the church of Whitson which is described by Hando as having a "thimble tower", similar to that of St Bride, but Pevsner calls it "a stone spike" with one of the corners higher, allegedly to stop it being drawn towards the sea. The only memorials there are to the local squires who are buried under a mound near the porch, which is now roofless so any trace of flood markers has been lost. Its history is poorly known, and there is no named saint's dedication. It was restored in the 19th and early 20th centuries but is sadly now used as a store for the local farmer. The updated Pevsner as recently as 2002 described it as "one of the most evocative places on the Gwent Levels, with the church and court nearby".

On the coast near Goldcliff are the remains of the Benedictine priory. Barber claims the first wall there was built by the Romans and that in hot summers the outlines of the old buildings can still be traced in the scorched grass. At low tide, the roots of trees show a forest grew south of the priory, reflecting Gerard of Wales' accounts, and similar was also unearthed across the Severn when Alexandra Dock was built. The region was also famous for shipwrecks, caused by nature and by wreckers; piracy and smuggling were also noted.

An anonymous broadsheet which described the Great Flood claimed an estimated "20 hundred" were saved from starvation and exposure by the boats carried on wains loaded with food and dry clothes. The tract ends with "The Lorde of his mereie[sic] graunt, that we may learne in time to be wise unto our own health and salvation, least that these water-flouds in particular prove but fore-runners unto some fearful calamities more generall."[13]

It seems the first sea defences were built by the Romans, probably to supply their forts at Caerleon and left a commemorative plaque by "Statorious" which was found in 1878. Romans built 20 miles of sea wall between Sudbrook and the River Rhymney, so it is likely they also fished there to add variety to their diets.

A prior and 12 Benedictine monks founded Goldcliff Priory from Bec Abbey in Normandy; it seems they founded the salmon fishery and built the Monks Ditch reen. But the monks were not wholly law-abiding. In 1334 the prior and a monk, and others from Newport, Nash, and Goldcliff, in Wales, and Clevedon and Portishead across the Severn, were charged with stealing goods including wine from a ship wrecked at Goldcliff.[14] By the 18th century the region was infamous for smugglers.

Along the coast past the Portland Grounds and slightly inland is St Thomas the Apostle at Redwick, the most interesting of the churches, though much of its history has been lost. The building bends north-wards from the nave axis, reflecting the Lord's head fallen to the north in death. It is part of a thriving community which allows the church to remain open. They have local history exhibits and visitors are often welcomed with organ music. The plaque on the outside wall is at a height of 5 feet, but Barber claims it covers a hole, raising questions as to the accuracy of its position. It states "GREAT FLOOD A.D. 1606". A lower one is on the buttress near the main gate.

The church is one of the best-preserved in the region, with a rare rood loft and screen. It is often open, so is the most visited, and also the most famous of the flood victims. It is part of a cluster of buildings with a bus shelter, which houses an open air museum, and a community centre. The museum makes no mention of fisheries, suggesting the industry had vanished by this time, or perhaps fish traps made of willow were at risk of damage or theft as the rest of the exhibits are more robust. The surprise inclusion is a cider press, as the drink is more likely to be associated with Herefordshire or Somerset. Carwyn Graves describes how Gwent was famous for its cider and orchards, of local varieties not known elsewhere. They were old trees, suited to the local, poorly drained alluvial soil, exposed to strong winds and salt spray. Slow growth produced intense flavours but made them unsuited to mass production. The apples and pears were grown for local consumption as cider and perry, especially for providing farm workers with fluid and much-needed nutrition during the long days of harvest. There is even a variety called Early St Brides which survives in the area. It is not known in England but is grown on the Continent, so yet another link via trade.[15] But the trees were not sources of profit so the government ordered most of them to be destroyed after the Second World War.

It is surrounded by meadows drained by reens with their own names such as Oxlease, Decoypool and Hare's Elver Pill, a reminder of the now-lost geography.

Magor is to the north east, claimed by Pevsner to be "one of the most ambitious churches in the country",[16] the earliest part of which is the 13$^{th}$-century tower. The piers are of West Country design, and one supports a green man corbel. This appears to be yet another church restored by Norton, so flood markers could possibly have been lost at his hand.

Undy is recorded in the 6$^{th}$ century and described by Barber a "a little brown church" and "a handsome structure"[17] until badly restored in 1878. He compared it with churches he saw on the Norfolk Broads. It has a 14$^{th}$-century bell made in Bristol, a rare reminder of the forgotten industry in the city. The vicar described how after floods, the water seemed to subside quickly, but pools sometimes remained for years, causing problems for farming, so these apparently stable areas of the region still have to deal with the ancient problem of floodwaters. The church dates from the 14th -century and its website claims it had a major restoration in 2001, and there is no mention of any flood markers. The vicar of Undy claimed that the floods were assumed to subside quickly, but some pools survived for years making problems for farming. The church is fortunate to be on a promontory of rock above the flood-level.[18]

A rarely noticed feature of the Norman font there is the presence of small cavities on opposite sides of it, used for salt and oil in the ancient baptisms, providing a sense of the region being untouched by the passage of time.

A church dedicated to Llanfihangel, St Michael, seems unlikely to be a victim of storms or flooding as the saint is generally associated with high places, especially hilltops, but he is also linked with rocky coasts. When seen from the moors, it is on a higher level, so close to the floodplain but was unlikely to have been damaged by the Great Flood. Barber notes its similarity to Whitson in having a "thimble" on a corner of the tower.[19]

In Portskewett, St Mary is a small, rubble-built Norman church without wealthy endowment. Its churchyard is boggy and overlooks a tiny stream, formerly its Roman port and ancient terminus for the Severn ferry. It was whitewashed inside in 1818 so any signs of the flood have been lost, but it has a monument to the Lysaght family of civil engineers.

St Pierre Pill and Sudbrook Pill were Roman harbours, with the crossing from the end of the Via Julia at Avonmouth was to either of them, both guarded by Sudbrook Camp.[20] They were thus likely to have suffered from the 1606/7 flood.

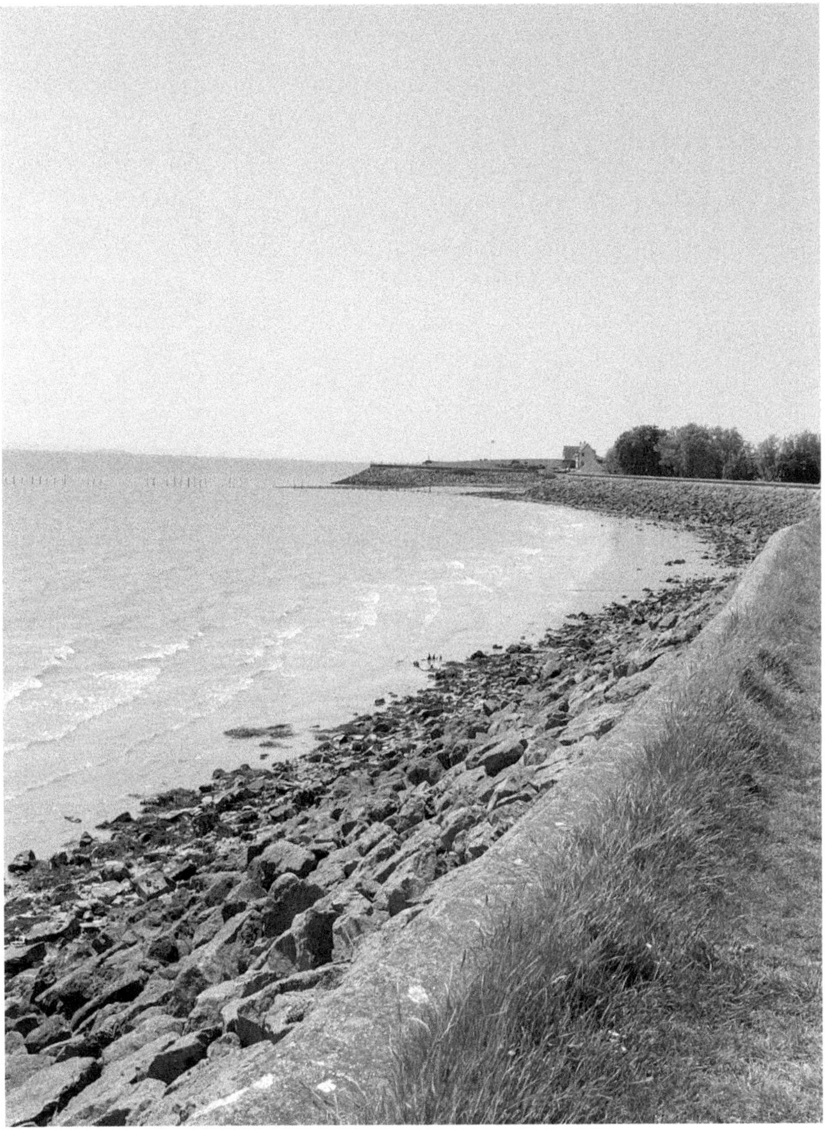

# CHAPTER 10
# WHY NOT WYE?

Of all the sites which are named as being damaged by the Great Flood, only Chepstow is on the River Wye, about 2 miles from its junction with the Severn. But this makes little sense as the river was on the line of the flood's path, and the region was industrialised from Roman times, so was well populated. Its southernmost point, Beachley Head, protruded into the Severn, so people there must have taken the full force of it. But it was isolated and rural, with much land lost since, and no records survive. It is now mostly a military base, so there are few people to preserve its history. It is opposite the Lancaut Peninsula and the landing stage for the Aust ferry. On the southernmost point beyond Beachley Point and Chapel Rock was the remains of St Twrog's chapel, claimed to be in England and its customs were controlled by Bristol. It did not have its own customs house to check ships' cargoes for taxation until the Sharpness Canal opened in 1827.

On the southern tip of St Tecla's Island, St Tylak's Chapel was recorded in 1290 but was probably much older. It had various names till the 16$^{th}$ century when it was recorded as being useless for being beneath the sea. Extraordinarily, in 1750 Ralph Allen of Bath proposed its rebuilding, but this was opposed by the site's owners, the Lewis family of St Pierre.[1] The riverside is now landscaped as part of its flood defences, as it is on the frontline of the river''s surges.

The dearth of detail on Chepstow is hard to fathom, as it was a major shipping and shipbuilding centre for many centuries, peaking during the Napoleonic Wars when most of the riverbank between Hardwick Cliffs and the bridge was occupied by timber yards. Its port on a floodplain, rising up to the church and castle on the rocky heights. The flood of 26 November 1703 was described as the Wye being in spate when it met "a rampant high spring tide", similar to descriptions elsewhere of the 1606/7 flood. But the same source claims the most destructive event was in 1969.

Romans crossed the Wye just above the town at Castleford, and a modern road runs across the bridge of 1816, allegedly built by John Rennie; it is 312 feet long, with 5 arches on strong stone piers. A bronze plaque, apparently undated, on the abutment clams the tide rose almost to its road level.[2]

Sherborne writes of how small ships, probably from Bristol, were paying tolls there as early as 1086.[3] In the 15th century, Bristol-born William of Worcester wrote of ships, mostly with food and timber, from various local ports docked at what is still called Welsh Back in Bristol's city centre. Chepstow was part of the same customs area as Frampton, Berkeley and Gloucester.

Mee writes of the half-mile-long town wall that encloses 90 acres in a loop of the Wye, running from the castle to a site facing Offa's Dyke on the other side of the river. He questions why much of the town outside the old wall was not built on, though there are suburbs.[4] This probably reflects the area being prone to flooding, so only workshops were built there.

Chepstow Bridge is claimed by Rhys to have been dangerous and wooden, "with planks so sprung and tenoned that in case of flood they lifted with the water". Crossing it on a rising tide felt like walking on stilts. It was well lit at night but sometimes the wind blew out the candle.[5]

An online source claims only 2 people drowned there in the Great Flood, which seems ludicrously low.[6] But the town is built around its castle on a rocky outcrop, with the floodplain beside the river being used as a port, and for shipbuilding and other maritime trades. The flood struck when people were up and about, so probably gave them time to escape to higher ground. By comparison the same source claims that at Arlingham, about 20 people were lost. This latter settlement is on a large loop of the River Severn, with no hills to see the approaching danger, and further from the safety of high ground. The same source clams Monmouthshire's total death toll was "only" 500, with thousands of cattle. Again, this seems low for an area claimed to be 39 by 6.5 km, with 26 parishes covered by up to 2 metres of water. Historian Rose Hewlett claims the fatalities were low, partly due to the fact that people were experienced in dealing with flooding. But this was no normal flood, the result of prolonged rain, and seems to have arrived with little or no warning, leaving little time for people to flee.

The Gwent Levels are made up of sedimentary layers from the last glacial period, between layers of peat from when sea levels fell. It is estimated that levels were then about 100 feet lower than now, but alluvial deposits have built up since then to form the salt marshes, with banks built to prevent the silt from reaching the sea. In areas such as the Somerset Levels, some areas are thus below sea level, and the silt deposits at the rivers' mouths prevent them draining. This instability is also shown in the rivers Wye and Severn which have frequently changed their courses. There are even claims that the Severn began as a tributary of the Wye, the latter of which has some of Britain's finest river cliffs.[7]

The narrowness of the Wye Valley makes it likely to have suffered high tidal surges, which may explain the images of churches and their towers being submerged by the flood. But records from the region are scant. The huge resource, the Victoria County History makes no mention of the 1606/7 flood there, though due to the numerous floods, many churches and their registers have been damaged or lost.

Kissack describes the early rise in industry on the Wye from Roman times, especially on its lower reaches, when wire works were founded at Tintern and Whitebrook, their mills driven by the force of water from the heights. But the plentiful, fast-running water often caused flooding and much destruction, especially to the many churches along its length. He cites extensive damage caused in 1735, 1795, 1852 and 1947.[8]

The large village of Glasbury, south west of Hay-on-Wye, had many "genteel houses" despite it being at high risk of flooding. The residents' wealth meant they could fund works to try to limit damage. In 1561 every manorial tenant living near the riverbank was ordered to plant 3 rows of osiers, alders or poplars along the bank to try and secure it. Defaulters were fined £5 for every rood unplanted.[9] But this failed, as by the mid-17th century the church and graves washed away, with the blame partly placed on the wickedness of parishioners.

Witney-on-Wye "has suffered more than most villages from Wye floods".[10] Three bridges were built and destroyed in quick succession before the 19th-century bridge was built. The great flood of 1735 was so forceful it created a new channel for the river, destroying the church and rectory so the church was rebuilt in 1740 on a different site with some of the same stones. Wordsworth died at the rectory. This is north of Hay-on-Wye and Kilvert often visited; it is now Grade II listed.

The town of Monmouth's risk of floods has been limited by modern urban sprawl, which allows much of the floodwaters to cover open fields.

Dixton is a small settlement north east of Monmouth, ominously between the rivers Wye and Monnow, so it should have been built on stilts. The parish is so small that Pevsner lists it as Alderton. It dates mostly from the 14th -century, and when I visited, parishioners were appealing for funds as they tried to dry it out from the latest inundation. Its churchyard is flat, and a gate opens to the riverside path to Monmouth. A local I met blamed the floods on trees being felled on nearby hillsides. Fortunately another source claims memorable floods were recorded on brass plates set into the nave wall. They include 10 December 1929 (1.63 metres), and 22 March 1947 (1.5 metres). This includes the apt "full fathom" with others in 1998, 2000 and 2002, so they seem to be increasing in frequency.[11] Pevsner claims the side chancel arch shows only the last 2 deep floods, with 1929 reaching 5 feet 2 inches and 1947 5 feet 11½. Floods are mentioned in the mid-18th & 19th centuries, and one in the 14th century. It seems to be one of the worst-sited churches and the dearth of records is exacerbated by Pevsner's claim that the interior has not been scraped. It must have been drowned in the Great Flood, but all trace has long been relegated in the procession of disasters that followed. He also claims the walls were free of damp, which seems unlikely.[12]

Hereford sits in a large loop of the Wye, so must have been hit by the big flood. Records cite October 1998 as being equal to 1947 which was only exceeded by the floods of 1795 and 1960. Its bridge was built in 1490 so was likely one of the first to be destroyed by the river.[13] There is no record of it being damaged in 1606/7.

The river Lugg is a tributary of the Wye, and is often flooded. A memorial in the church describes a flood caused by a great storm "which produced a wall of water 20 feet deep", destroying buildings and gardens, drowning 4 people.[14]

Whitchurch is another riverside church which is also often flooded. In the churchyard are the graves of the Ballinger brothers who drowned in 1853.[15]

It seems the absence of records for the Great Flood is thus largely explained in the region by it being obscure, the buildings since destroyed, and by the many floods which have happened since.

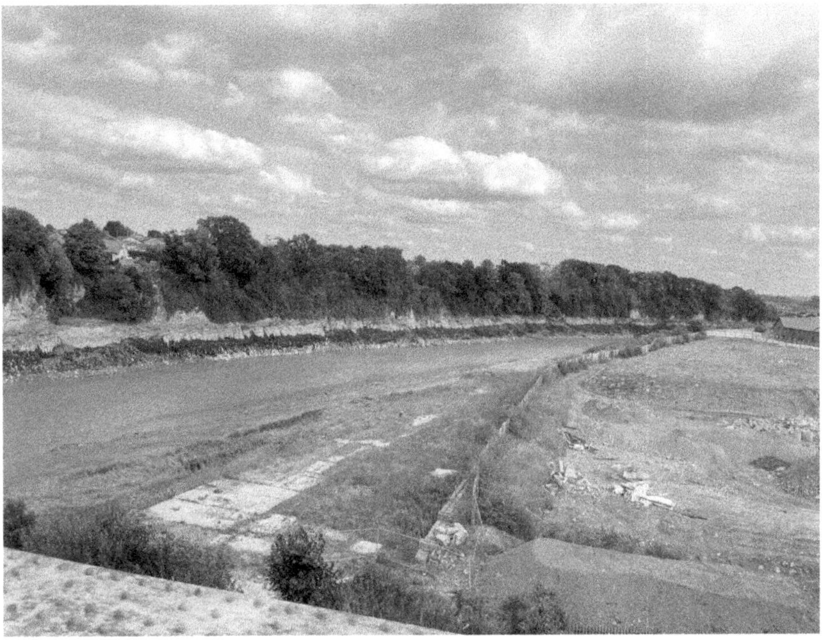

# CHAPTER 11
# SEVERN TIDELANDS

The River Severn begins on Plynlimon, a mountain in Powys; it originally flowed north to Cheshire and the River Dee. But in the last Ice Age it was blocked by huge rocks which flooded Cheshire, causing the river to turn south into its present course.[1] Smith and Ralph describe how the Severn was formed when, in the last Ice Age, its waters broke through the cliffs at Ironbridge in Shropshire, changing its route from north into the Dee to the south. This source also cites floods causing havoc in 1607, 1957 and in 1770 when a record flood of over 34 feet hit Gloucester.[2] Below Worcester, the Severn mostly flows through "flat green meadows, or "hams", past Upton-on-Severn and into Gloucestershire near Tewkesbury.[3] Beyond Holt Fleet, the river is prone to high floods; the hams become lakes, and the river hard to navigate. By the time Worcester was settled, the sea had receded, leaving swamps and the uplands were well forested. The former swamps became commons in the south of the county where some areas retain the term 'marsh' to reflect their past.

Just over a mile of water can be seen at low tide as you look from Avonmouth to the Welsh coast 5 miles away. For about 25 miles downstream of the bend at Fretherne, the banks are a mix of silt and sand banks.[4]

Like Bristol's Avon, the Severn has one of the highest tidal ranges in the world, and is also susceptible to dangerous floods, "causing havoc in 1607, 1947 and 1770 when its highest flood tide was recorded at Gloucester of 34 foot".[5] The term 'spring tide' is often mentioned in relation to the river, as it peaks in spring and autumn, but it is not strictly related to the season. It happens when a tide surges up a river to meet runoff, causing a rebound in the weaker force. This was common at Tewkesbury, especially in the 13$^{th}$ century, and is now increasing again, putting the region at high risk of floods.

Possibly the best definition of the region is that of a 'poured landscape', the result of glaciers withdrawing, allowing seas to replace them, producing a ridge of clay on sand along the coast, and layers of silt inland. The River Parrett, in Somerset's flatlands, deposited layers of silt by its annual floods. But when it reaches the sea it is opposed by the twice-daily tidal surges of the Bristol Channel. Instead of the traditional notion of a river flowing to the sea, the fresh water is forced back inland, sometimes causing its natural flow to halt. Thus there are parallels with the famous tides of the Nile, bringing fertile soil to the floodplains which allowed the expansion of early civilisations. Without such waterborne fertility, animals were needed for their dung to be spread onto the fields. But these plains were short of wood for buildings and for fires, so locals often used animal dung, mixed with straw, for heating and cooking. The need for sea walls and irrigation canals gave rise to early civil engineering, and to much of what we recognise as the modern world. Such large-scale engineering works also forced people to co-operate on a large scale, to plan, to experiment, which in turn allowed populations to expand, but also led to conflict and wars.

The Severn Channel is funnel-shaped, so the incoming waves are squeezed, which concentrates their energy and creates higher tides when their force should be diminishing. But this gives them minimal reflected energy with strong bottom friction. This is not so very different to the Thames and many other river estuaries where the force has dissipated before it enters the main river. However, the Severn's height increases as it heads east, making its height at Newport 60% higher than at Tenby. Sloping water in the Bristol Channel creates a horizontal pressure gradient on the downslopes. It produces big spring tides: where motorway bridges cross it can be 14 metres/24 feet, i.e. as high as the walls of Caernarfon Castle, and 13.6 metres at Newport's peak. When strong winds blow, they create round pressure systems, creating the Coriolis effect, rather than pressure moving from high to low pressures.

The estuary area is about 10 km$^2$, with a tidal range of 5 metres which involves 50 million m$^3$ of water in the 6-hour tidal influx, an average of 2,300 m$^3$ per second.

The Severn is famous for its bi-annual tidal surges known as bores, which drill into the main river, transporting eels and fish upstream faster than the main river, so acting as a hydraulic pump. The direction of the river flow immediately reverses when the bore arrives, and this is often announced by a low rumble, and seen as a steep wall of disturbed water surging upstream. This allows the Severn to be surfed for several miles.[6] But the bore is not guaranteed to occur, as heavy rain can prevent it. It moves at faster than walking speed, and is inversely proportional to the width of the river, so it moves faster in shallow water. The main bore is also associated with 'undular bores', called whelps.

If the bore reaches a bend in a river, it can be carried on by whelps, reflecting off the bank and reducing the bore to chaos. It may not happen at all in wet weather. Though the Severn's is the most famous, partly due to its proximity to settlements including Bristol, the phenomenon in a weaker form is also recorded on the Taff in Carmarthenshire, Usk in Gwent and occasionally near Swansea on the Lougher Estuary, and at Bridgwater.

When Romans first saw the bore near Gloucester they believed the world was about to end, so their response was similar to those who witnessed the 1606/7 flood. There are 250–60 bores each year but they are only big enough to attract spectators at the spring and autumn equinoxes. They occur during spring tides: named not for the season but when the upcoming surge meets the downflow of the river, causing the lower to spring back. They happen in clusters of 3–5 days, so attract sightseers and surfers. The tide is at its highest between Beachley and Aust where the height can be 14.5 metres, only exceeded where the riverbed is lowest at the Severn Suspension Bridge as the tide enters the Severn and it narrows from 100 miles wide to less than 5.

Big spring tides are common near equinoxes, i.e. spring and autumn, and are more significant in dry weather. Big spring tides at equinoxes have also been observed on the Trent. Storm surges are often caused by low atmospheric pressure coinciding with strong onshore winds, which can overwhelm onshore defences, especially on the west coast.

Trawling through old records, there are mentions of land being swamped, washed away, eroded or reclaimed, especially at Fretherne, Longney, Arlingham and Hill where the Romans carried out works. Medieval records survive from Fretherne, Saul, Frampton-on-Severn and Slimbridge. At low tide, trunks of old oaks can be seen at Alvington.[7] It seems the salt marshes to the north of Bristol at Henbury were drained in the late 12th century, which allowed the expansion of agriculture to feed the rising population during the Medieval Warm Period. As this was virgin soil, it was initially very productive, so the area became a profitable agricultural region. But this came at a price, as from Roman times the risk of flooding continued and nutrients in the soil were depleted.

Surviving records provide a strong sense that the Severn's banks have always been in flux, causing problems for travellers, and disputes over lands, especially at the water's edge. Mention is often made of land being eroded, stolen, replaced or sliced through. Such events were generally described as acts of God, and happening in 'time out of mind'. But it was increasingly due to the actions of men, for failing to maintain sea walls or clear drains, so they survive in the written records as they became matters for the local courts. Sometimes these incidents were the aftermaths of storms, but they provide a sense of local people forced to live with, and adjust to, uncertainty. It was increasingly common following the Reformation, which caused the loss of monastic engineering skills and the role of the church in planning and enforcing storm defences and drainage.

The area between the New Passage and Severn Beach was especially at risk, as the river was forced to turn north there, so must have taken the full force of the 1606/7 flood. It probably came from the south west, so likely had a shearing effect along the Welsh coast, with main force aimed at this region. But strangely the present walls appear to be lower than those on the Welsh side. Materials for building and repairing banks were mostly sourced locally to save effort and transport costs. But if stone and mud were dug too close to the defences, this risked forming pools inside the defences which undermined the walls and delayed post-storm drainage.

With no central government to fund and supervise these engineering works, the size and quality of defences varied. It may also have been hard to obtain local workmen, as the best time for construction coincided with the busiest time in agriculture, in fair weather. Engineering works in cold, wet weather were demanding, dangerous and time-consuming. With the closure of the monasteries, the engineering skills and the ability to mobilise large workforces was also a large substantial problem, and largely left to individual landlords to solve.

Romans began land reclamation of the estuary in Gloucester, and evidence survives at Elmore, Longney, Arlingham and Hill. In the medieval period, there were also works at Fretherne, Saul, Frampton-on-Severn and Slimbridge. Henbury's marshes seem to have been drained in the 12$^{th}$ century, which coincides with the expansion of Bristol's port, so the region provided food for the growing city and for export. When lands were drained, they were very fertile but declined over time unless managed floods continued to supply waterborne silt. By the 13$^{th}$ century Romney Marshes in Essex had been managed 'time out of mind' with enforcement for those who failed to maintain their ditches and defences. Thus the famous 'wooden walls' of England, i.e. its armed ships, were preceded and have been outlived by its walls of stone and mud.

The term 'sewers' entered common usage in the early 16$^{th}$ century, suggesting the weather was changing, or perhaps reflecting the closure of the monasteries with their engineering skills, and/or perhaps the climate changed so more drainage was needed. The first statute of sewers was passed in 1531/2.[8] The first record of commissioners included Stephen de Salso Marisco for part of Henbury's Salt Marsh.[9] So the region was not just a flood pain, and must have been of value for some time to have become a family name. Was this also a region of salt production, an important industry, especially for the Romans, but which seldom leaves any historic traces?

Surveys were made of the sea walls from at least 1410, so the river's misbehaviour was of long standing by then. The use of fish weirs further complicated the situation, causing silting up of the waterways and at times their removal was ordered.

The Severn is infamous for its unpredictability, which added to the problems in trying to contain it. Significant floods were recorded on spring tides in 1687/8, 13 February 1688/9 and 27 November 1703, when boats travelled over hedges. Following the 1606/7 flood, orders were made to increase the heights of walls, including part of the wall at Slimbridge, which was ordered to be raised by 2 feet. The area has been a major victim of floods, with the whole area impassable in the early 17$^{th}$ century. A house was destroyed there and the tenant allowed to build a replacement on higher ground. Hewlett notes complaints against the tidal mills at Oldbury which were often blamed for storing water which then flooded the surrounding fields. Similar disputes are reported before the Reformation near Salisbury when the monastic mills were often blamed for flooding farmers' fields and obstructing drainage. There were also conflicts between fishermen whose racks of baskets anchored in the river interfered with shipping, and some were ordered to be removed.

Hewlett mentions a chantry certificate from Henbury concerning a pre-Reformation legacy to fund a priest's prayers for the donor's soul. But when the chantries were closed, the funds were diverted to repair the sea walls. She also mentions a flood from 60 years earlier — possibly the Duke of Buckingham's Water — when 220 people drowned including the Duke. She claims the parish was large and poor so unable to otherwise fund the works.

Arlingham also diverted chantry funds to maintain sea walls, tide gates and trenches against flood tides.[10]

Hewlett also mentions repairs at Hill in August 1636 as being just in time for the flood the following 4 November. It seems it was similar to 1606: a high spring tide and a strong south west wind leading to flooded cellars in Bristol, Clevedon and the lowlands between Bristol and Aust.[11]

At Elmore a bushel of wheat funded repairs from 1555, and John Guise, lord of the manor, left a legacy of 13s 4d in 1588. This shows that while the size and extent of the Great Flood were unprecedented, it struck in the midst of many other extreme weather events. There had clearly been problems for decades in maintaining the flood defences. But the closure of the monasteries in 1536-41 caused a loss of engineering skills and of large-scale, ongoing management to carry out necessary design, maintenance and repairs. If this was true in Gloucestershire, the situation must have been widespread across the Severn region.

Bristol often claims to have the highest tidal range on earth, second only to the Bay of Fundy in Canada. But high tide can reach 48 feet at Chepstow. Kissack claims the annual rainfall at Plynlimon, the source of the Usk, Wye and Severn rivers, is 90 inches, reducing to ¼ of this at the famously floodable Tewkesbury, on the Severn.

Romans built fords at Forden and Caersws, and locals must have been constantly dealing with the Severn's misbehaviour. Churches built flood walls at Minsterworth and Arlingham, though to little effect. The earliest recorded disaster dates from 18 October 1483 or 4 when the Duke of Buckingham rebelled against his cousin Richard III. Following 10 days of torrential rain, the Severn burst its banks and the king expected to trap Buckingham on the west bank. The duke might have taken the safer route via the Westgate causeway, but it was already under water. Gloucester was flooded, houses damaged and children floated about in their cradles. The water remained high for 10 days. This seems to be recorded by an 8-feet-high mark in Tewkesbury friary. Buckingham was caught and executed on 2 November 1484, with claims that he should have been left to drown, rather than hanged.

It seems the region was spared for a time, but disasters returned in late 16th century. Kissack claims the walls of Worcester record high flood levels in 1672, 1730, 1770 and 1795. The last seems to have caused the most harm as it is recorded by a plaque stating "This plate is fixed to let you know/that Severn to this line did flow". Telford claimed the weather was unprecedented, with frost and snow building up before a sudden thaw, causing the biggest flood in English history. Bridges were damaged by lumps of ice bashing against them, with 16 destroyed in Shropshire alone. Disasters continued into the 19th century when the floor of Minsterworth church was raised by 4 feet to protect it.[12] Thus it seems that the great age of British engineering was the result of many disasters which needed such skills.

Throughout its history the Severn has thus been a source of food, a means of transport, but its floods meant it also posed risks to life and property. Until the construction of turnpike roads and the improvements made by engineers such as Macadam, most traffic was by water. Riverside lands were valued for grazing and fishing. Land near Gloucester borough and castle was lush meadows, with "bridges, causeways, fish-weirs mills, warrens". The Severn meandered and the bore eroded the river banks downstream, to deposit upstream, often causing blockages to navigation. There is evidence that the meanders have been stationary at Elmore since Roman times. From the mid-17th century, the middle of the stream at Alney was an island. But by 2006 the floodplain in Gloucester was described as "a semi-derelict landscape of rubbish tips, trailer parks, wetland nature reserved and disused roads".

In 1535 the king ordered the destruction of all fish weirs on the Severn, but a fish house survives from 1610.[13] Both fish weirs were destroyed by the Gloucester Commissioners for Sewers, as they were blamed for silting up the navigable waterways.

Over Bridge was rebuilt in 1591 with stone from Ashleworth, but like many others, it was probably damaged by the Great Flood of 1606/7, as it was rebuilt in 1611 for £121, of which £22 was spent on stone.[14]

Robinson claims in 1914 that few parishes in Gloucestershire could claim greater antiquity or a more picturesque appearance than the village of Henbury. "[It] includes in its area several outlying villages and hamlets, in a very large parish, and is said to have derived the first portion of its name from the word Haen, meaning old. It is in the hundred formerly called Birnintreu, and in very early records is described as "Henbury in Salso Marisco" [salt marshes] ... Evidence of

Roman occupation is plentiful in the parish."[15] When the Great Flood hit the region, the Mayor of Bristol organised boats to rescue victims from the Aust salt marshes.

In Lewis Wilshire's book 'The Vale of Berkeley', published in 1954[16], he describes Hallen as being doomed, as the area and that of Henbury were both on the brink of development. A farmer called Clem Hymell claimed that 68 years previously, the Severn had reached Hallen where a new house was being built, and where many pigs and cows drowned. The flood also allowed the salt to ruin the wells and local water sources.

A mere 40 years earlier, the village of Severn Beach was isolated by floods for 3 weeks per year. Clem described how these dangers were balanced by the abundance of free food, of the fishing industry between Avonmouth and the Aust Passage. Locals hunted rabbits, birds, pigeons and foxes, and used ferrets to kill rats, with bounties often paid to control these vermin. There were plentiful sprats, the surplus of which farmers carted to Bristol. This allowed local people on low incomes to survive before the welfare state. But all this was ended by the construction of Avonmouth Docks.

Clem claimed "most of the land adjoining the estuary locally is marsh. Without rhines it would be unproductive, so needed constant upkeep." Historically, drains and sea walls were maintained by HM Commissioners of Sewers via great local landlords, i.e. Berkeley, Poyntz, Chandos and Throckmorton. Significantly, from c.1600 there is an appeal for help:

"Whereas the countrie after the Great Flood cam downe into John Hortts orchard in Redwick, (Gloucestershire) and there cut a great slewes (sluice) to let forth the water. Never since made up sufficiently to defend the same, but every high tide your poor supplicants, being dwellers there, bee greatley damnified and almost un donne thereby."

The repair costs were £42, which seems a lot in the early 17th century, especially as most of the locals were living on seasonal work. Thus as the varied forms of local subsistence went into decline and population and rents fell, the great landlords were failing to deal with the increasing problems of flooding on the low-lying coastal area, making it increasingly unviable for settlement.

William Adams also adds confusion by claiming there was a flood at 9pm on 4 November 1636 "the south west wind was blowing very hard on a full spring tide" sounds very similar to the 1606/7 event. In Bristol, shops and cellars on The Back were filled, sea walls were overwhelmed in and about Kingston and Clevedon, the low marshy grounds between Bristol and Aust and various other places suffered great harm, and many beasts drowned, "Yet this flood in my judgement, and by the judgement of others that set marks for both, do affirm, that it was not so high by a foot as it was in the last great flood in January 1606.[17] Thus it seems that marking the heights was a common practice, so the dearth of such records reflects the absence of gentlemen interested in such matters in the parish at the time of the Great Flood, of subsequent neglect and/or removals.

At the time of the Great Flood, the Royal Commission of Sewers had expired, so a new one was needed to organise relief and recovery, from Hungroad to Longney and Minsterworth. The problem was worsened by the deaths of some of the commissioners, and others had moved away. It seems the main blame for the flooding was placed on the pair of tide mills at Oldbury which were accused of letting in salt water, preventing the outflow of fresh water. This caused significant problems to local people, many of whom were poor. Claims were made that 200 acres were damaged by the damming up of fresh water and especially the entry of salt.[18] Thus, there were many ongoing problems in the area before the flood amplified the neglect. In an orchard at Redwick, south Gloucestershire, a sluice was cut to drain the waters after the flood, which cost £4, but it was not repaired, which raised fears of more high tides.

In April 1607, i.e. the same year as the Great Flood, concerns were raised of the risk of sea floods between Kingroad at the mouth of the Avon and Berkeley and adjacent parishes needed protection. Claims were made that walls were too low, and that grain was ruined by standing in salt water. This is different to other reports which mentioned stacks of grain, so this is referring to winter crops, suggesting the output of the region was high, meaning the losses were likewise. The huge amount of at least £10,000 is mentioned as lost. Lady Mary Stafford owned Oldbury Mills which were blamed for storing water 2 feet higher than the surrounds, so when the storm hit, this flooded 2,000 acres at Oldbury, Kinton, Morton and Rockhampton. Commissioners ordered the gates to be lowered by 2 feet and repaired, under threat of a £10 fine. A range of work also included repairs to sea walls and drains.

On 2 October 1607, a meeting was held at Thornbury to discuss the late tempest between Hungroad andMinsterworth where "divers people perished", a worryingly vague number.

In the parish register of Rockingham near Berkeley is the following: "on the 20th day of 1606, the sea did overflow the banks and sea walls, insomuch that very many people and cattle were drowned all along by Severnside from Bristol to Gloucester". There is a tradition that waters rose as high as Frampton Tower, at least 60 feet above the present level of the Severn".[19] Though several images and accounts mention incredibly high flood waters, this is the only one to provide a specific place and height.

Before the establishment of modern local government, landowners were expected to manage local affairs, and in the wake of the Great Flood, they set to work with impressive promptness. The flood was in mid-winter, which put the region at risk of further damage. Records of the Gloucestershire Court of Sewers (drains) survive from the time, and provide a valuable insight into how badly the region was affected, and what steps were taken to repair damage to waterways, walls and, in particular, mills, the latter of which were often complained of for depriving farmers of water, and of letting dams overflow, so harming homes and lands downstream, as well as letting floodwaters bypass flood defences.

The footpath along the sea wall from New Passage to Aust was described as "typical of the estuary". The New Grounds/Warth beyond the new walls were mostly used for cattle grazing, and the presence of seaweed there shows the floods continued. This is the area of coarser grass "on ground recently reclaimed from the sea", which is noted as "still underwater in heavy tides" with many wildflowers showing the battle with the sea was ongoing. This is a reminder that the boundaries between the river and dry land were in a constant state of flux, which is hard to imagine in our modern age.

Problems with erosion continue near Gloucester, such as at Stonebench, a popular site for watching the Severn Bore.

We tend to see floods now as disasters, especially when they cause so much damage to property and to lives, but annual flooding from the days of Egypt on was welcomed as providing valuable nutrients for the soil, so avoiding the need for animal manuring. The dire state of roads until recent times also meant that the major form of transport was by water, of making the Wye and Severn vital to many people. This caused conflict between locals who built weirs for fishing, and boats which needed clear waterways, especially when water levels were low. But if the water was too high, vessels could not safely pass under the increasing numbers of new bridges. Telford claimed that in many parts of the Severn there was only about 1 foot of water in 1796, the year after another great flood. This must be the same event which Kissack refers to at Newton when Penson rebuilt the church and the bridge which "replaced a wooden bridge which had surprisingly withstood the 1795 flood".[20] Montford village was "some distance" from its eponymous bridge, a predecessor of which had been there in the 13th century when 4 pence was paid for each ton (barrel) of wine or honey, and 1 penny for each Jew on horseback, reduced to ½ penny if on foot. But it was not always safe, as a messenger for Edward III was drowned "by the rising flood of water", and could not be found, so that he was "devoured by beasts".[21]

But extreme weather was not only about floods; Upper Arley Iron Bridge was built in 1861, the longest cast-iron bridge at the time. Its designer, Telford claimed it was completed in a single season to take advantage of the low water of the drought.[22]

The Severn Estuary has long been a busy trade route. But it was always a difficult and dangerous navigation, often shallow and famous for changing its direction and routes. Waters described a local man as having an almost mystical relationship with it, constantly watching its ever-changing moods. He knew it as well as anyone, but not as a friend, but as a beast that could turn on him with little warning. The river provided a huge amount and variety of fish for locals to trade and to eat, especially salmon and lampreys, and into the 20th century, there was still a huge export trade in elvers to Germany. But it was always difficult and dangerous to navigate, often shallow and prone to change direction.

Wilshire claims: "Of all the villages of the vale Framilode alone really belongs to the Severn. Church and Inn join with people in looking across the water that has figured so largely in their lives. Cider ad salmon together form the foundation upon which these places were built, but orchards can be trusted to keep their places while the river needs watching... there could be no pleasanter spot than the bench beside the inn from which you can look out on a scene that is not disturbed by the ripple of a wavelet or the cry of a gull."[23]

Smith and Ralph claim in 1172 the river washed away a large piece of land at Slimbridge, moving it to the opposite side at Awre. Then in c.1570 it eroded the Awre bank, returning the soil to its previous site. This reflects claims that what is now the Wetlands Centre was once the centre of the river. John Smith of Nibley wrote that it eroded the Awre bank, "byn cast togeather and left by the River of Severne". He recalled it had been known by him when the grounds were the body of the main part ... and the deepest part of the channel.[24] Its site on the inner bend of a large loop of the river meant its graveyard received many victims of drowning. But this danger was compensated for by the rich soil for pasture and orchards, supplemented by its salmon fisheries so the dangers were seen as acceptable. In the parish registers, a Francis Browne was drowned, and buried on 12 January 1606/7, so a rare mention of a victim of the 1606/7 flood.

In 1234 Awre on the west bank again lost some of its land to Slimbridge and went to law to retrieve it. But this failed as the jury testified the disputed ground had originally been pasture owned by Slimbridge; it was only a temporary gift to them from the Severn so it had been returned to its original owners in 1639.[25]

Fisheries were a major source of employment and food, but it is unclear why they died out. Some locals blamed pollution at the ports, others to land reclamation through the introduction of sea grass by Lord Fust of Hill which turned the salmon away, as they like to see the bottom of the river. Much has changed in the region; industries have moved in and traditional practices been marginalised and lost, though traditional fisheries still continue in limited numbers, under government licenses.

Another flood victim may have been John Fryer of Cross Lane Arlingham who was buried on 8 February 1606/7. It may have taken some time to find his body or the delay may have been due to him succumbing after some time to hypothermia. Francis Lockyer was drowned and buried on 12 January 1606/7. Dates are clarified in the register, with a William who drowned on 19 January 1606/7 (the day before the Great Flood) but was only buried on 27 March 1607, suggesting the level of chaos after the disaster.

Lord Berkeley's private army built Arlingham's sea walls during a pause in the Hundred Years' War. Waters described it as "the work of fortification for the protection of the realm against the forces of nature... well suited to soldiers unskilled and indolent at husbandry".[26] The castle's location on a floodplain seems a bad choice, but the region provided good grazing, and could be deliberately flooded to drown invaders. Charles I became fond of salt-fed lamb, setting a court fashion which expanded grazing on land beyond the sea walls, so encouraging reclamation of lands there.

Yet despite their strength, the walls were still vulnerable. Waters describes a recent but undated breach 40 yards wide, which isolated locals, who were forced to "move upstairs with their pigs and poultry". Oddly, the region had no boats, so the local innkeeper took his spring cart to rescue his neighbours. Waters also described the river as a thieving cat, often depositing strange "gifts". A small whale was seen at Awre in July 1943, the first of its kind in over 20 years. The beast returned on the tide but died, and locals used part of it for pig feed. Its solitude suggests it was sick or weak.[27] Fixed nets were used for fishing on the Severn Estuary to capitalise on its huge tidal range driving fish into the nets, then trapping them as the waters receded. But this was impossible below below Avonmouth due to the fierce currents, and where more robust basketwork traps were used there. Salmon were fished from large intertidal pools and lakes left by the receding tides at Oldbury-on-Severn where the salmon pool was c.2 miles long, and at the upper end was 15 feet or more wide, and 3 or 4 foot deep where it narrowed down to the weir. The salmon were caught in nets in the muddy water, which travelled with the current as the salmon tried to escape back to the river. But fishing was only possible at low tide as water travels so fast and with such great force.[28]

Conical openwork baskets made of local willow of c.6 feet long and 2 feet wide were lashed in rows at right angles to the river, so salmon were forced in by the rising tides. Their heads became stuck, so they were unable to reverse against the current. Yet yields were low – of 650 putchers they only obtained 6 fish during the season of 15 April–15 August. Any other animals were caught, even sharks and Porpoises. Kypes were also used to catch large amounts of shrimp and small fish. These were used in the Bristol district at Hallen, Oldbury-on-Severn, Shepperdine and off Berkeley. Nets were bag-shaped and hung on the foreshore where fish entered on the ebb tide, and emptied at low water. Nets were 12–18 feet square, up to 30 feet long, and the catch was mostly sprats, whiting and flounders.[29]

Waters writes of how the Severn curves away from the sea wall below Berkeley Pill, where the ebb tide exposes a large area of wasteland at Hills Flats. The Ordnance Survey map shows the river there to be a dangerous, complex mix of shoals and waterways. But as with other so-called wastes, it was farmed as intensively as the lands within the walls. Large basket fisheries were fixed to the river bottom as weirs, known as 'fixed engines'.[30]

Hill has already been mentioned and has several further claims to fame, being one of many sites where Joseph of Arimathea planted his staff. Lights were maintained there for safe navigation, which explains the presence of a hermit's chapel there. It was funded by Thomas Lord Berkeley in the 14th century when storms were common both here and in the Low Countries. It is directly opposite Woolaston chapel, a landmark for sailors. This may have attracted the benevolent Fust to the area to drain it, as the site was described as "evil in winter, grievous in summer, and never good... The chapel is now(1947) cottages, the only place where seaweed was used for manure.[31]

Hill is to the south of Berkeley, described as being flat and sparsely populated, and seems to be little more than a church in an estate, rising above the surrounding countryside. The parish church is appropriately named after St Michael on the hill, and is largely a mausoleum to the Fust family.

Trinity House records that the church was whitewashed as a landmark for shipping, but its spire was removed after the great storm of 1703. Much of its interior was destroyed by a fire in 1897, possibly removing more details on the family which barely rate a mention in the Dictionary of National Biography.

Nine pills (tidal streams) were recorded between Slimbridge and Littleton parishes, each of which were forced to maintain its own share of flood banks, which must have been a huge expense to these regions. Bigland notes a churchyard stone commemorating a boy who was shipwrecked and lost with the captain and 2 other seamen in a violent storm of 1735. There were probably many others whose records have not survived.

The old Roman road separates the the inland parishes of Aylburton, Alvington and Woolaston, which were rebuilt in the 19th century and so have no traces of earlier storms.

Epney had a major eel fishery, but elvers were tiny, c.1,000 per pound. Germans ran a collecting station to ships, sending c.7 million per annum across Germany. They were attracted with flares and caught, so the river resembled a city street. Locals ate them, and pressed them into dried 'cake'.

By 1941 the fisheries were much reduced, but they still included flying fish and seahorses as food for the poor, though large lampreys and sturgeon were very rare.[32] The diversity of food before the war is shown by mentions of post-war Bristolians still eating 'wall fish', Helix apsersa, the common garden snail.

Brian Waters wondered why the church and a house across the road from it at Tirley, were the only ones in the village at risk of regular flooding. The reason was that the land was very flat, so in summer the river was ¾ of a mile away, but winter floods reached the churchyard, spectacularly in 1924, as recorded by a hole in the church tower which was within 6 inches of the mark for the great flood of 1852. The church used to have a paddock, and its wildflower-laden grass was mowed to be strewn on the floor of the church every Whit Monday. But the location of the meadow is now lost, "dwindled and swindled away".[33]

On a suitably overcast day in September I drove south along the coast of South Gloucestershire to try to imagine the region in the past, especially the troubles local people had had with the water levels. Northwick Tower's fine ancient animal heads are similar to those seen at the Norman church of Deerhurst near Tewkesbury. A rooster perches at an angle on the top, the grounds below a billiard table flat surface of well-tended turf. But the main church has been lost, and the tower itself, fenced off as its last rebuilding, is now at risk of collapse. A variety of monuments survive, and burials continue. It seems there was a church at Northwick in the late 11th century which was ruined by 1370. Was this due to neglect, poor building skills, or more storm damage?

The church had fallen into decay, and was badly restored in 1842. "The tower fell, and was rebuilt in the west end, the restoration being hopelessly bad. In 1913, on all Saints' Day, the church was completely restored at a cost of £700, and rededicated to the service of God by Dr Browne, Bishop of Bristol. There is no record of the past history of this church but the parish registers give an account of serious inundations of the Severn from time to time, similar to that which occurred in December, 1910, when nearly every house in Northwick had the river water through it."[34]

It was part of the parish of Henbury, one of the largest in the West Country, comprising 4 churches and some 10,000 acres, as well as the hamlets of Kingsweston, Lawrence Weston, Bowden Fields and Upton St Leonards.

The site doesn't warrant its own chapter in the voluminous 'West Country Churches', but it is mentioned in the chapter on Aust, the ancient church claimed to have been "a place of considerable historic fame" as it was a common site from which to cross the Severn. It can be seen on the famous poster of Bob Dylan for the film 'No Direction Home' where he was at Aust ferry with the first Severn Bridge visible under construction in the distance. Romans shipped goods and men from here across to Caerleon and Caewent. Aust is also famed for its scenery, as from the top of Aust Cliff the monument to William Tyndale can be seen at Nibley Knoll. The churchyard has several memorials to people who drowned nearby, but the strong tides mean that many bodies were never recovered, which was likely the fate of many people who were lost in the wake of the 1606/7 flood.

None of the above is particularly helpful, but fortunately some information boards at Pilning and Severn Beach provide some valuable insights on the history of the region.

They begin by describing the region as "essentially marshy, often inundated", suggesting the population was mostly sparse and poor. They claim boats were probably launched from Pill Head and some drainage works may have been carried out, but the present landscape was not established till the early 17th century, which sounds about the time of the Great Flood. This in turn raises questions as to whether the first defences were in response to this disaster or to the generally turbulent era. The article claims the marshes were a deterrent to building large settlements, with Redwick and Northwick being outpaced by Pilning, and especially Severn Beach in recent decades, though this is still not saying much. Pilning is likely named from Welsh 'pyl', a pool or creek. The region is still largely used for light industry, with flood defences adding to the cost and risks of new buildings. In Saxon times, Redwick was the place of reeds and 'wic' meant dairy farm. The absence of sheep reflects their susceptibility to foot rot.

It seems there was a church at Northwick in the late 11th century which was ruined by 1370. Was this due to neglect, poor building skills or storm damage that drove the expansion and fortifications of inland settlements of Bristol and Norwich?

Further south, at Weston-super-Mare, fish stalls between Birkbeck Island and the mainland peaked in 1830–80 and were mostly lost when the pier was built. The Island had a hut for "gull yellers" to scare gulls away. There were usually 5–6 tons of sprats per catch, which were sent to Bristol and Bath.

Following the eastern side of the estuary, here are details of possible flood sites:

Littleton-on-Severn is separated from the Severn by orchards and meadows, so it was low-lying enough to have been drowned. The church has beautiful old tiles round the font's base and on the floor of the tower which are from Thornbury Castle, some with the arms of the tragic Duke of Buckingham of Water fame. The following details come from Mee except where stated otherwise.

Beachley is a little village with a quaint landing stage for the Severn ferry boat, so must have been swamped.

St Thecla has a holy well and hermit's house of prayer which survives but only as an arch from the chapel which was abandoned 400 years ago.

Tidenham is on the peninsula between the Severn and Wye, its medieval church a landmark for shipping, and the 600-year-old church has fine views of the levels, so it escaped flooding, and likely provided shelter for the survivors.

Lancaut is on a bend of the Wye, a place of "almost unbelievable beauty" with green meadows and the sparkling river below, but now only a roofless old church.

Alvington is on the highway, close to the Severn. When the tide is out, trunks of old oaks are visible.

Aylburton is about a mile from Lydney, on the line of the old Roman and medieval road. Most of the slopes were wooded and the Severn bank was about 1km closer to the main road as about half the current "levels" were reclaimed before 450AD and cleared of trees by the 14th century. Aylburton Pill was still used for shipping in 1608. Though there are no records of storm damage, it must have suffered.

Lydney is on a long switchback on the road between the Severn Estuary and the Forest of Dean in a beautiful setting. Mee describes it as follows: "when the wind blows up the wide cliff-edged estuary, rippling its water into waves, we might almost fancy ourselves by the sea; we may even see a wheat-laden ship from Vancouver calling at Sharpness opposite, the furtherest inland port in England ... There are 2 hills separated by a valley about 30m wide"[35] which would have allowed floodwaters to reach inland.

Awre is one of the most famous places to view the Severn Bore, so was likely damaged by the Great Flood. The church is famous for a chest claimed to be over 1,000 years old, roughly hewn outside and inside with a Saxon adze. It is huge, at 8 feet long and 3 wide, claimed to have been used as the village mortuary where bodies were laid before burial, so it must have suffered badly from the Great Flood, but the records are lost amidst so many other tragedies.

Newnham-on-Severn is on the outer edge of a large jigsaw-shaped bend, but it has been burnt several times so records of 1606/7 don't survive. Described as in "a lovely situation on the bank of the Severn", it has a fine stretch of sand by the river and a street sloping gently up to the church on the cliff. But by the 14$^{th}$ century the river was undermining it so was dismantled and rebuilt on higher ground close to its present site. This in turn was destroyed by fire in 1881.[36] The church bell chimes the tune 'Home Sweet Home'.

Twice last century the church was refashioned, the second time after a fire in 1882, but it has kept a few relics of the past. These include "a fine Norman font which came from the old church at Nab's End before the Severn covered up its site."[37] Facing east, it has extensive views across the Vale of Berkeley, the Severn, and the Cotswolds. During the Civil War, Royalists were garrisoned there, and a barrel of gunpowder exploded in the church, killing a score and damaging the building. The local history website claims it was originally settled due to its prosperous river-borne trade and the convenient river crossing there. A crossing was probably in use from the 1st century. A ferry is recorded from 1238 and had extensive trade with Bristol and Ireland till the opening of the Gloucester Ship Canal in 1827. In the late 18$^{th}$ century the ferry was described as very safe, and carried horses and carriages. But it seems this partly involved the use of a long jetty from Arlingham where passengers were carried across on men's backs. This jetty likely interfered with the Severn currents. Several attempts were made to build a tunnel and a bridge, but the muddy base and strong currents made this unviable. Adam the Fleming witnessed a deed in 1220, and about the same time several Lombards were there, with Adam trading with Santiago, suggesting another refugee from floods in northern Europe.[38]

Epney lies upstream on the same grand loop of the Severn, but to the east. It is a tiny hamlet famous on the Continent for its elver trade as already mentioned. It has been inhabited since the late 13$^{th}$ century, so was part of the Norman trade expansion.

Arlingham is east of the wide, low-lying loop of the Severn, 13 miles south of Gloucester with a road leading to a ferry a mile to the west. The church looks beyond the river up to the Forest Of Dean.

Saul is south, in the middle of the loop, "set in a lovely place where the Severn makes a loop of 9 miles to cover 1". It has a tiny church with several memorials to victims of drownings in its churchyard.

Next to the south is the large village of Frampton-on-Severn, close to the river's edge and between several large lakes, one of which has a sailing club. Its 22-acre green is one of the country's largest; it includes 3 ponds and a chestnut grove. It was a marsh, reclaimed by Richard Clutterbuck who built his home, Frampton Court, nearby. At the end of the village, among the trees and beside the canal is the church.

Slimbridge is the next place to the south; its main claim to fame now is due to its popular Slimbridge Wetlands Centre, formerly called the New Grounds as its existence is due to the Severn shifting west from the centre of the parish. The Wetlands Centre was established due to the area being so flood-prone, making it unprofitable for farming. Nearby is the vicarage surrounded by a water-filled moat.

Next stop south is Purton, 3 miles south of Berkeley; the Gloucester and Sharpness Canal passes through it. By 1282, a ferry crossed the river to another settlement of the same name, suggesting it began as a single village with a stream through it which expanded into the main river. This raises questions over settlements such as Redwick, found on opposite sides of the river, which may also have begun as a single settlement but been forced apart by rising river levels. The ferry at Purton Passage, was made redundant when the Severn Railway Bridge was built in 1879. It is now protected from tidal erosion by a barrier formed by several hulks which were run aground at high tide, which have since filled with silt. A new decoy pond was built in 1840 on the New Grounds i.e. reclaimed land, between the canal and the Severn.

The Sharpness Canal was built to avoid the delays to shipping caused by the Severn's tidal swings, especially the danger of running aground when the water was low. It was the broadest and longest canal in the world at 26.5 km.

Oldbury-on-Severn is best known for its proximity to the Oldbury nuclear power station. The village has 2 churches and many orchards at the end of a creek. It provides fine views in all directions, with the

Welsh mountains beyond the Severn. The Romans built a defensive camp there, and earthworks known as the Toot, the old Saxon name for a lookout. The church is on a small hill, so a fine viewpoint and waymark for travellers. It was almost destroyed by fire in 1892, which may explain the lack of flood records there.[39]

Littleton-upon-Severn is to the west of Thornbury near the mouth of the Severn, but the parish has merged with Aust. Its church, with an unusual dedication to St Mary of Malmesbury dates from the 14th century but is largely a rebuild of 1878. Nearby is a chain of reed beds, another marker of flood risk, close to the river and managed by Slimbridge for migrating birds to rest and feed.

| To Drain this Parish | from its Owing Flood |
| To Model and Repair | this Howse of God |
| Are Paterns good & fit | to Future Time |
| Free Profit yours | & Costs Labour Mine |

S.r Francis Bull Baronet
Lord of this Manor for
y.e Benefit of its Inhabitants
at his own Expence Plan'd
Built and Erected in the
Year 1760 the Great Sewer
at Hill Pill Next y.e River
Call'd the Imperial Drought
and y.e two Others above it.
He also in 1756 New Modled
and Repair'd this Church.
All the Costs and Materials
for all the Said Works
are a Gift from him
Freely to this Parish for
the Use Above.

Let those of Ability Strive to do Good

# CHAPTER 12
# SEVERN CROSSINGS

One of the most iconic photos in the history of popular music is that of Bob Dylan on the poster for Scorsese's film 'No Direction Home' where he is at Aust awaiting the ferry to Wales. The First Severn Crossing can be seen under construction in the background. The bridge begins where Aust Rock intrudes into the river, touches base at Great Ulverstone rock, then flies over Beachley with its army barracks to land in South Wales. The old ferry had a landing pier near the Hen and Chickens outcrop. William of Worcester, the mid 15th century "painstaking student of English architecture" and history is recorded as returning from a visit to Tinterm Abbey on the Wye. "where he tells us that he dined with the brethren and landed near Aust"[1]

Historic sources mention people crossing the Severn on foot or on horseback, especially during the Civil War, but I've spoken to locals who claim this would be madness now, so the river must have changed. In the reign of Henry VIII a pirate named William Chick was the terror of the Bristol Channel, described as its "chief Champion"[2] the Star Chamber court claims he was part of a group of pirates who travelled from Bristol to Cardiff via Aust Passage. Though Ordnance Survey maps show there are extensive patches of rock and gravel, there is no obvious safe path. The river is notoriously muddy and therefore slippery, so anyone taking this route would have to be either very desperate or foolhardy.

Higher upstream, many river crossings prevented the river from becoming a national boundary. Romans crossed its upper reaches to reach the Forest of Dean's mines and a temple dedicated to the Celtic god Nodens survives at Lydney.[3]

The lowest Roman bridge crossing on the Severn was at Gloucester, so they settled there. Maisemore is 4 km north west from it on the west bank of the Severn. Hutton writes of an ancient, high-pitched bridge over the Severn, the repair of which was funded by people leaving legacies in their wills. The Commonwealth rebuilt it. He adds that he crossed the river twice, along the causeway, then using Telford's Bridge to Over where the Romans had a camp to defend the river crossing.[4] This was the original, Roman, land route to Wales.

In my trawling of old, long out-of-print books, I found an intriguing, and fairly recent route, describing Roman roads in Gloucestershire. A tiny pocket book, 'Topographical and Statistical Description of the County of Gloucester' by G.A Cooke Esq is a wealth of valuable details. It lists gentlemen's — and a surprising number of ladies' — estates, stagecoach inns, fairs and bankers. There is a general description of the county and of the Via Julia from Bath across the Severn into Monmouthshire via Durdham Downs, now part of north Bristol. It states "between King's Weston Inn and the mansion house, descends between that and the stables, and passes straight by Madan farm, till it joins the banks of the Severn".

Then comes a bombshell: "Here was a ford into Wales; and part of the road on the opposite side of the river to Caerwent existed, still paved, a few years ago."[5] This is confirmed by another source which claims that a Roman road led to the "old ford" across the Severn, which remained visible and was still in use till about 1802. The most recent mention of the ford seems to be a reference to the Bristol riots of 1831. Bishop Gray preached at Bristol Cathedral on the Sunday morning but was forced to flee, "followed by a wild mob" so sought shelter at Knole Park, Almondsbury. He had intended to cross into Wales via the "Old Passage".[6]

Ernest Rhys provides a dramatic account of the sea crossing in the early 20th century at Easter when "it was blowing and raining hard. The Severn was rough and dirty enough for any open sea, tumbling and rolling with choppy muddy yellow billows before a west wind. It was more daunting than the Holyhead itself in the teeth of a gale — the flooded Severn about an hour after high water, looked malevolent. The wind, too, in the station at Portskewett — crying, buffeting, howling, and whistling — was such as only Dickens with his uncanny faculty of describing the elements at odds with roof and walls, could describe."[7]

Yet from Roman times, ferry crossings ran from Aust; on maps it was called the Old Passage, as previously mentioned. A steamboat ran from New Passage but by the 1830s faster boats sailed from Aust again. In 1886 Brunel's railway tunnel made ferries obsolete.

One account claims Prince Rupert was chased over the "English Stones", which again raises questions about a road across the river, but it seems more likely he fled for his life with his horse swimming part of the way.

Fred Hando claims both St Pierre and Sudbrook pills were still tidal in the Roman era, and the Severn was crossed from Avonmouth in England to St Pierre Pill or to Sudbrook Camp.[8] The latter was still joined to the Gloucestershire side by a sandbank which was exposed, so passable, at low tides. This was used during the Civil War when a group of Royalists forced locals to carry them across, but they were left there to drown by the high tide.

A ford or ferry into Wales may explain why there are Redwicks and several other settlements on both sides of the Severn.

One of the accounts of the the Great Flood, 'God's Warning to His People of England ...' includes: "At Aust, many passengers that are ferried over there now, are seen to be led by guides, all along the canals, where the water still remains for the space of 3 or 4 miles, or else they will be in great danger of drowning, the Water still lying so deep there."

Cook provides more detail on the Severn with "the tide, well known in the Severn for its boisterous and impetuous roar, comes up to Gloucester with great rapidity and violence, and turns the stream as high as Tewkesbury. The tide generally rises 7 ½ feet at Gloucester. At Framilode Passage the saline impregnation begins to be lost."[9]

Claudius' invasion in 43 AD tried to hold the Cotswolds and Severn Crossing with defensive earthworks, bypassing the Belgic River crossing at Arlingham-Newnham, forcing the Romans to cross at Gloucester.[10]

Arlingham is now in the centre of a large loop of the Severn where the land often shifted. But HPR Finbergh refers to its ferry being a driftway at low tide, formed by the movement of water or ice. It was a drove road for cattle at low tide over the Severn at Newnham, through which the main road from Gloucester to Cardiff ran to supply the London markets. In 1171 Henry II gathered troops at Newnham for the invasion of Ireland.[11]

The first Arlingham-Newnham ferry is recorded in 1238. By the 18th century it was described as a safe river crossing, till mudflats made the access unsafe following the Second World War, probably due to its use and damage by heavy vehicles. Severn trows had flat bottoms, so were suited to the shallow parts of the river, but they still needed high tides to reach Gloucester. Arlingham was the site of a bizarre drowning 200 years ago when a fisherman was holding between his teeth a plaice he had caught, to free his hands from the net. But the fish jumped down his throat, choking the man till he collapsed and drowned.[12] A more cheerful tale from the same source is: "the Severn Bargees used to wait for high water in Arlingham, coming ashore, in a state of maudlin inebriety," drinking the money that was meant to support their families, so they would end up raiding local farms for fruit and vegetables. Attempts were made to build a tunnel as an alternative, but the mud made this impossible. A bridge was also rejected.

Bristol historian Reverend Seyer described the Severn during the 1703 storm, saying the area "suffered deeply on the Forest of Dean side, but nothing in comparison of the other side, from about (H)Arlingham down to the mouth of the Avon, particularly from Aust Cliffe to the river's mouth (about 8 miles), all that flat, called the Marsh, was drowned. They lost many sheep and cattle."[13]

Newnham-on-Severn is on the outer arc of the huge Severn loop around Arlingham. The original settlement was at risk from erosion so moved to higher ground in 1366. A book on the town by M.K. Wood[14] claims there were several violent floods, and in 1607 — i.e. the Great Flood — the banks were overflown 6 miles on either side. Rudder claimed in 1779 it was a crossing for wagons and people on horseback at low water, "with more reputation than prudence," as many lost their lives in the attempt. Hence any claims that a river crossing existed at a certain point do not mean it was a safe one.

Downstream from Newnham is Awre which is claimed to have changed little since the 1200s. In the parish church is an enormous chest, an ancient 'dugout' claimed to have been used for laying out the bodies retrieved from the Severn.

The Haw Bridge crossed the Severn where the floodplain was 3 miles wide, with Hasfield some distance to its west. Wikipedia claims "as much of its land resides below the 50-foot contour, it is subject to regular flooding". Downstream is Ashleworth, a fine farm with a manor house of the 15$^{th}$ century, tithe barn and church facing the village green. Wikipedia claims "The oldest part of the village is Ashleworth Quay, on a floodplain on the west bank of the River Severn." The church has suffered frequent floods, sometimes reaching higher than the pews, with a wooden plaque showing the water level from the 1947 flood about a foot above them. The masonry has survived for some 900 years. Across the river is the straggling village of Sandhurst which a source claims failed to develop further due to to floods.

The river narrows as it passes Minsterworth, making traffic difficult at low tides. The ferry and its customers were often at risk of floods, hence the church is dedicated to St Peter, the fisherman.

Epney was famous for its elver fisheries, and live elvers were shipped to Germany to stock their rivers until the Second Word War; 4 million were sent in 1908. The parish had other industries such as fine local cider and teazles for weavers.

Bigland claims there was an alabaster industry at Aust, with blocks thrown up by the tide which could at times be very violent. He also noted that the force of the tide sometimes led to increases and losses of land beside the river, which made crossing and navigating the river difficult and dangerous. On the tops of the cliff large teeth of ancient animals were also found, which may have added to the instability of the land.

Alvington was called Abone by Camden, a name shared by Sea Mills on the Bristol Avon. Claims were made there about "the remains of oak trees, lying with their roots to the north east after the soil they grew in was washed away by the encroaching tide", again echoing Gerald of Wales. He added that as recently as the Roman settlement, the river was "not more than one fourth its present width" and "it is indeed curious how suddenly it narrows to about that width above Awre".[15]

The Victoria County History describes salt marshes were being reclaimed at Lydney by the early 18th century, suggesting the river is now narrower than at the time of the flood. It continues: "most of the broad tract of level ground to the south of the parish has been reclaimed". By the mid-16th century a thin strip of land included the salt pill, and 80 acres of the New Grounds, but they had been washed away by the early 19th century.

There was a Severn crossing from Purton hamlet to its namesake in Berkeley parish in use by 13th century, which suggests another site where the settlements may have had the same origin and been driven apart by a stream which became part of the main river."[16]

The peninsula and river mouth were part of Gwent. A small chapel was founded at its most southerly point, which is now separate, again reflecting the turbulence of the waters. The chapel was popular in the 14th century. Chapel Rock was allegedly home to a hermit, which would make sense, as hermits maintained roads and bridges, gave help to travellers and buried the dead. A chapel dedicated to St Twrog was built, possibly to maintain a light for navigation of the river, which was the role of hermits. But it had decayed before the 18th century. It probably extended further into the river in antiquity, so was an early, or perhaps the original, crossing point. The chapel was popular in the 14th century

The peninsula later had 3 inns, and Defoe visited Aust but failed to cross due to the weather, so detoured via Gloucester. The alternative crossing, New Passage, was between Redwick and Pilbury in the east to Sudbrook.

Reverend Samuel Seyer of Bristol provides an extraordinary claim that from 20 November 1607 — apparently the same year as the Great Flood — a severe frost caused the rivers Severn and Wye between Bristol and Gloucester to freeze over for 9 weeks. This allowed people to travel from one side to the other, played gambols and build fires to roast meat, echoing the Elizabethan Frost Fairs on the Thames. But this also meant that no boats could reach Bristol, threatening the city with food shortages. Seyer also wrote of the Great Winter of 1683/4 when seas froze within 2 miles of the shore. In the Great Winter of 1635 there was high snowfall and ice on the Severn which prevented Bristol Fair and lingered a further month, so about the same time of the year as the 1606/7 flood.[17]

Romans settled around — but not in — Bristol. There was a road from Bath via the northern suburbs to Sal Mills, now Sea Mills, a port on the Avon, 3 miles inland from the river mouth. A ferry travelled from there to various ports on the Monmouth bank of the Severn where a scapula of silver and coins have been found which date from 49–52 AD. This route was likely used to avoid the difficult terrain of the Forest of Dean and lower Wye. The port was probably linked to Bath and involved in Julius Frontinus' involvement in South Wales from 75 AD when Caerleon-upon-Usk was settled as the frontline and Caerwent settled soon after.

But it seems odd that Sea Mills would be a crossing to the settlement in Caerleon. Mendip lead was shipped from Portbury to the south, so what was its role? The Roman presence may be linked to a path north from the Avon via Henbury and the salt marshes where the inhabitants sought shelter from the Great Flood by climbing trees. Salt was a major industry, but one which seldom leaves any trace. A passing mention of women collecting crystals which formed on the ground may explain its invisibility. The Welsh word for salt is Halen, which is found on maps of the area. Was a salt industry the reason for the settlement?

The River Wye joins the Severn at the narrow headland of Beachley beneath the original Severn Bridge. The area has been eroded over time, so probably reached further south and east. It is tidal as far as Redbrook. Monmouth is 13 miles upstream with steep sided valleys which channel heavy rainfall to feed streams that drove early mills. But these same conditions caused the region to be frequently flooded, so the 1606/7 Flood was likely seen as the most extreme among many others. The oldest industrial site was Kings Mill, of 1434, which was a major port. Yet it only rated a single mention in a book on the Port of Bristol in the Middle Ages when their ships had paid tolls from 1086.[18] Today it is described as a nice village with some pubs, and only rates a single paragraph in Pevsner, for its Victorian parish church with nice tiles, so there are no records of the flood.

The centuries of human activity in the region converge at the 2,000-year-old Sudbury Camp, close to the 700-year-old, heavily eroded Trinity Chapel. Not far away at Caldicot is the entrance to the Severn Tunnel. It has a pumping station which became an unexpected expense when Brunel's railway tunnel flooded when diggers damaged a spring. The project was in part funded by sales of the spring water. It still needs constant pumping to keep the trains running.

Human activity further declined at Beachley Head as by 1758 locals were no longer mowing grass there, though the churchyard was still in use for burials when Captain Blethin Smith of Sudbrook was carried to his grave by 6 seafarers in the late 18th century.[19] Yet despite the erosion to the church, the local stone was valued for its ruggedness and was used for the lower part of the Newport bridge, but this quarrying likely hastened the site's decline.[20]

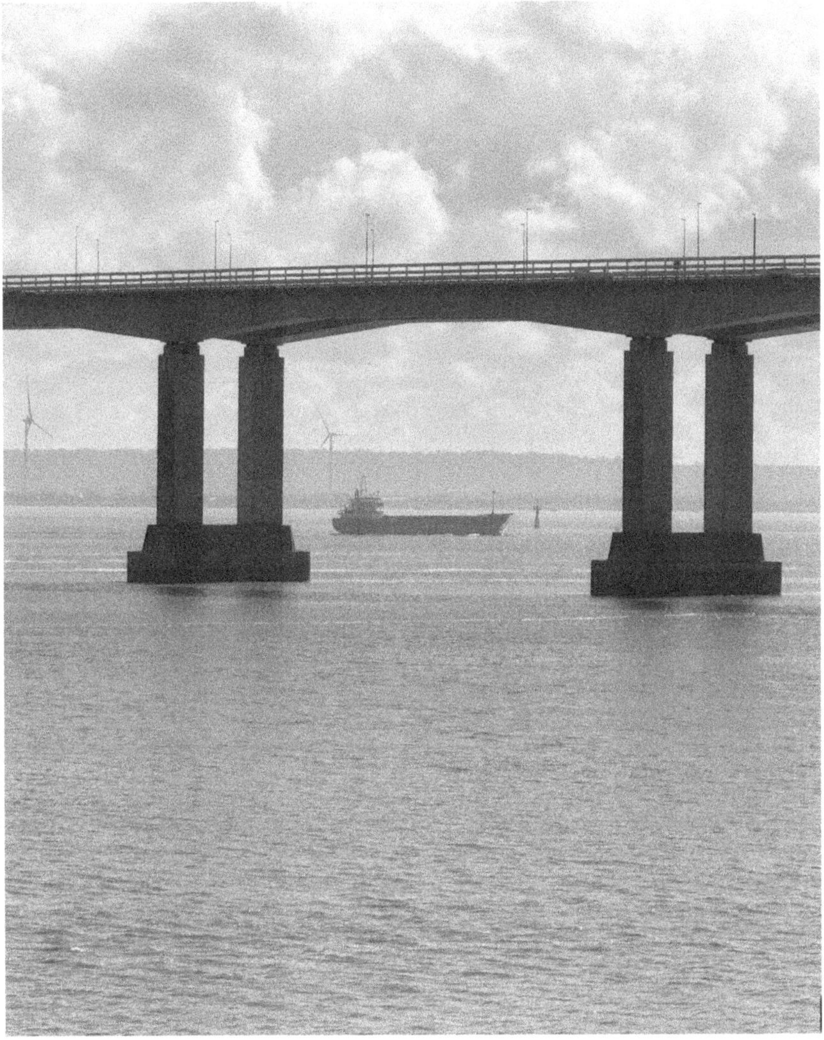

# CHAPTER 13
# GLOUCESTER STORMS

Bigland describes a vigil on the Day of St John the Baptist — 24 June — 1258, when "an immense inundation ensued through heavy rains and all the shores of the Severn, from Shrewsbury to Bristol, were so flooded that several adults and innumerable children and animals were drowned."[1] This was the period when the North Sea was infamous for its huge storms and flooding.

The only record of the 1606/7 Flood from a local, traceable source comes from a report in the parish records by John Pawle, the vicar of Almondsbury, several kilometres inland, dated 26 January 1606, 6 days after the disaster. He claimed the tidal wave had surged up the channel from Minehead to Slimbridge which affected the Severn, but also the Wye, Avon and even the Monnow.[2] This is the the sole mention of the latter river. He wrote: "in Saltmarsh many howses overthrowne. In Hobbes house syx foot hyghe. In Ellenhurst at Wades the sea rose neere 7 foote and in some howses there yt ran yn at the wyndow and out at the other. At Chepstow 2 drowned and Arlingham, an infamouse site of floods, about 20 were lost."

In our modern age it seems that whenever heavy rain is reported, the press features images of Tewkesbury surrounded by water, which suggests the site is an unsuitable one for a building as large as the abbey. But the abbey itself is rarely underwater. The only marker I found there dates from 22 July 2007, a few inches above the floor. Like many ancient religious houses, it is on flat land, surrounded by waterways which generally keep the worst of the floodwater at bay. Before the construction of surfaced roads in the 18th century, water was the main means of transport, crucial to trade and for driving mills to provide food and industry. Tewkesbury has suffered many floods, with the Victoria County History listing them in 1484, 1587 and 1611 but surprisingly not 1606/7, though the river froze hard enough the following year to be able to bear the weight of people and wagons.[3] The lower Severn also froze and people walked across it, as described by Reverend Seyer previously. It notes that 1652 and 1653 were memorable for having no floods in summer or winter, showing they were so common to be considered normal.

Various woodblocks of the 1606/7 flood show water as high as church towers, which seems outrageous, but in narrow valleys this might have been true. In the Forest of Dean, especially where slopes had been deforested for industry, there may have been a perfect storm with fast-flowing streams meeting the incoming tide to cause incredibly high floods. Extracts from the Berkeley family archives describe how the sea overflowed banks and sea walls at Rockhampton and that many people and cattle were drowned along the Severn from Bristol to Gloucester, with one claim that it rose nearly up to the top of Frampton Tower, over 60 feet above the Severn. Tandy adds that there was a normal high tide and "ferocious gale" that lasted 5 hours, a rare mention of wind and unique record of the event's duration.

The storm struck at a time of extreme weather when it seems the winters had become very harsh and Lord Berkeley had already organised a commission into the state of the Severn's banks. Some of the commissioners must have lived close to the coast as they are recorded as having been drowned by the floods. The people of Berkeley had already petitioned for the repair of the sea walls. The separate tower of Berkeley parish church was destroyed by either by this flood, or that of 1703, which shows the chaos of the age.

Rockhampton is on a reen of the same name, north of Thornbury, inland from Oldbury-on-Severn. Tandy claims the sea overflowed the banks and the sea wall, drowning many people and cattle there. This was repeated at many sites, as the sea was high enough to flow over their sea walls and banks, but was then unable to drain away until new breaches were made.[4]

After the Great Flood, the whole of the Severn Valley was described as a disaster area, with live animals clinging to dead ones. Milkmaids were said to have been overcome by the flood whilst working. The flood was said to have lasted 7 days, meaning that is how long it took for the walls to be broken to allow the waters to drain away.

At Minsterworth the flood was described as a "thorough tempest", suggesting rain was involved, "whereas of late, great hurt and damage of the rage and overflowing of the sea that happened in the county of Gloucestershire".

John Smythe who worked for the Berkeley family left records of other floods: "In January, in the 4th year of James, AD 1610 was so great a tyde, made greater with a strong winde that it overflowed a great part of the lower county adjoining, drowning diverse men, women and cattle of all sorts." But James I reigned from 1603, so the 4th year was 1607; allowing for the change of the calendar, this seems to be the Great Flood again. Smythe adds that a similar event had happened 20 years before, and was already "growne remarkable in many histories".[5] This suggests a date of 1586, 2 years before the Armada was sunk, and Smythe adds that flood prevention works had been built at Berkeley Pill, with dams built "at every inlet of river and stream". This seems to be the only mention of this earlier storm, though there are references scattered through accounts of defences being overwhelmed.

Ships sailed up Berkeley Pill (tidal stream) to dock at the bottom half of the High Street, a reminder that the area had no bay for its port. Many coastal manors included ships moorings, generally on pills. What we now see as a largely agricultural region was thus heavily involved in long-distance international trade, and many local gents invested in early colonies, as reflected in the names they gave to settlements they founded in what became the USA. The diarist Parson Woodforde claimed there was a tidal range of about 10 feet at Pill. The lower part near the future power station would have frozen. The tide rose to 28 feet at Berkeley, flooding the castle, and the waters reached 18 feet at Newport, south Wales. Alkington was about the same height as Berkeley, so should have been safe from the Great Flood, and Woodford was on high ground, so not a flood risk.

The adjoining parish of Ham saw the Severn overflow "the like of not being before heard of" as it rose above the tops of the houses, causing major damage. Another source claims the lower lands near the Severn "were frequently overflowed by the tides, so must have been a morass". This in turn reflects the long-term changes in use. The site of the castle was chosen for its boggy ground as a form of defence, and open space gave plenty of warning of enemies approaching. Lord Berkeley's troops were put to work between the wars to repair the sea walls, with a commissioner to monitor the jurors and committees. Thus it was a major, ongoing task to maintain the region once its defensive role became redundant.

The author lists high floods in the Severn Valley in 1561, 1565, 1606, 1620, 1628, 1689, 1703 and several in 1900.[6] Tandy wrote a book of his childhood in the 1940s when flooding was common, and the river froze up to 1 foot deep in some winters. Some places, such as the Salt House, flooded annually, but the worst in living memory is claimed to have been 1947, which was a combination of snow melt, high rainfall and high tides. These were followed by extreme events in 1965, 2000 and 2001.

He claims the biggest of them all was 1607, so he has converted it to the modern date. It covered the biggest area, up to Worcester, the only mention of this city, and remained for 7 days.

The number of deaths caused by the Great Flood and the many others ranges widely, with one author claiming there were none, which is ridiculous. Searching local registers shows no clear pattern, as each parish varied widely in the damage caused, and the responses and rescue efforts. In Ham, the flood was "the like of which had not before heard of, and that it was above the tops of the houses". Cause of death was not always listed, which made the situation more complex, due to the high levels of plague and of food shortages, in these turbulent times.

The extent of the floods must have varied widely in respect of the slope of the land, the width of channels, speed of the water etc. A book on Newnham-on-Severn claimed there were "several violent flows" in

1607, when the river was said to have overflowed the banks on either side there.[7]

The 12th century historian William of Malmesbury described the Severn as the county's broadest in its channel, "more violent in its flows... in the daily fury of its waters, sweeps the sands from the bottom and piling them up... but it does not extend further than the bridge of Gloucester". This latter comment seems to reflect the state of London Bridge, which had been repaired so many times that it was more a weir than a passage for water. The famous Frost Fairs were common in 17th century London, but the Thames upstream from the bridge was also wide and shallow as it was not embanked.

Another source claims that between Newnham and Gloucester are many lost grounds, at risk of the Severn's sudden rising, and that at times the entire district of Arlingham was a marsh.

All of the above shows, as in previous chapters, that the Great Flood did not come out of nowhere; it was the biggest of many extreme weather events, and its local impact seems to have varied widely.

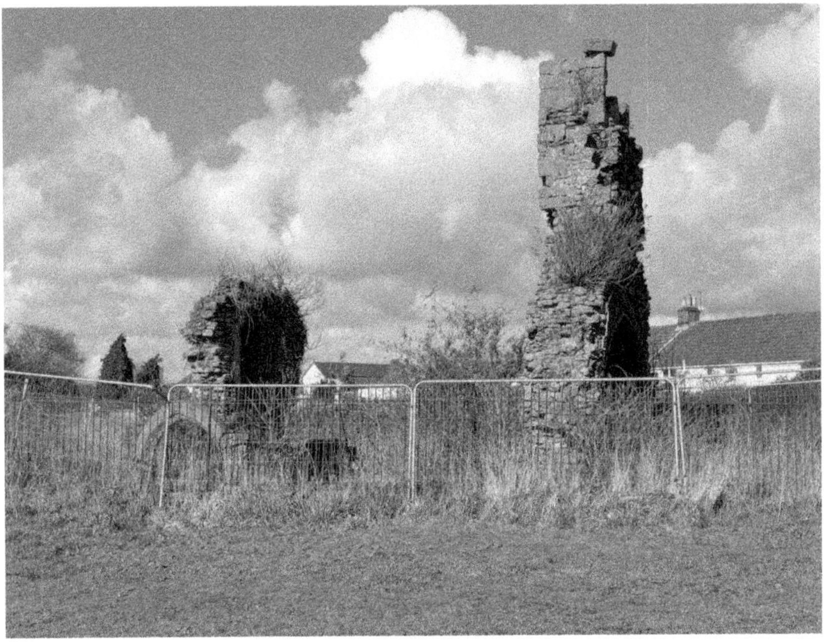

## CHAPTER 14
## BRISTOL AND ITS HINTERLAND

The best-known source for the Great Flood is Jones' 'God's Warning to his People of England...', but it was apparently written in Wales and the region's largest settlement, Bristol, scarcely rated a mention. The city is 8 miles upstream on the winding River Avon, so the force of the flood must have been diminished by the long journey and by being forced into the many creeks and inlets along the way. But the storm struck at fair time, so the city was full of traders from London, Wales and Ireland, meaning there were more people and goods to be harmed.

From the above source comes: "In Bristol was much harm done, by the overflowing of the waters, but not so much as in other places." The castle and much of the city centre were on high ground, beyond the reach of the flood, though many buildings had double and triple cellars into which the waters may have entered. Some merchants' houses were along the water's edge, with moorings for boats to unload direct into their cellars so they were easily flooded.

Jones continues with: "Many cellars and warehouses, where great store of merchandise was in, (as wine, salt, hops, spices, and other such like ware) were all spoiled. And the people of the town were forced to be carried in Boats, up and down the city about their business in the fair time there."

Bristol was in many ways a special case, as it was the largest and most populous city and port in the area with the rivers Frome and Avon running through its centre. But the long, winding River Avon with its many creeks to dissipate the force should have provided it with protection. In the 13th century, it was improved by a massive investment in engineering works which diverted the River Frome, rebuilt the town walls further out and rebuilt the eponymous bridge, allowing the northern Gloucestershire and southern Somerset parishes to be united. The south is mostly low-lying so at higher risk of floods, though it was also home to the fabulously wealthy medieval Canynge family. Fears of Spanish invasion led them to improve their maps, which highlighted weaknesses on the coast at Aust, Oldbury and Shepperdine.

The city's recorder, Ricart wrote in October 1484 of ships lost at Kingroad due to "the greatest floode and greatest wind that was ever seen... Nearby salt marshes were flooded, cattle drowned and houses swept away by the sea". In Elizabeth's reign, a survey claimed between the city and the coast there were an astounding 59 "pills, creeks and harbours" for smugglers, some of whom were city merchants".[1] As investment in ships was often shared to spread the risk, this must have involved many city residents.

This is from Latimer's Sixteenth-Century Bristol: "In 1565 the Common Council learnt with consternation that an effort was being made by the inhabitants of Gloucester, then a mere 'creek' of Bristol, to procure an independent Customs House for that port. Petitions against a proposal regarded as highly injurious to local commerce were hurriedly despatched to London, the Lord Treasurer's aid was besought with a "gratification", and the rejection of the project was temporarily secured. ... But in 1580, to local dismay, Queen Elizabeth, by letters patent, established a Custom House and attached to it the upper creeks of the Severn. Earnest protests against this were addressed to the Privy Council, who, in 1582, directed a Commission to sit at Berkeley to inquire into the merits of the case"[2]. The matter was so important that Bristol raised a special rate on its citizens to fund their legal challenge, as the city claimed it had owned all the creeks on the Severn between Berkeley and Worcester "time out of mind". Most of the city's food arrived via the Severn from as far as Shrewsbury.

Bristol claimed the loss of the customs fees threatened the city with bankruptcy. They also claimed much of the Severn trade was in small boats which paid no customs charges, so were cheating the Crown, and that their own trade was in such decline that many of their ships had been sold.[3] But the request was granted in 1580 to cover the northern 'creeks' of the River Severn.

Bristol was also the centre of a wealthy hinterland at the time of the flood, so it was in the city's interest to launch rescue boats, and it provided emergency food and dry clothing. The city was home to wealthy merchants on the corporation and the now-infamous Merchant Venturers, who had the means to organise relief and support. Fortunately, the dearth of detail from other sources is compensated by the wealth of information recorded by the Reverend Samuel Seyer. His detailed records suggest the flood of 1606/7 was part of an unsettled period, which may explain why records elsewhere have been lost. On 4 October 1604, he claimed, "was the greatest snow that was ever known by the memory of man, which continued for 4 days. And by reason that the leaves were upon the trees, very many were thrown down by the roots, and the limbs of many were broken into pieces."[4] This was serious as wood was put to so many uses, for house and shipbuilding, furniture, wagons, buckets and firewood.

His description of the Great Flood is worth quoting in full, as it has been cited by other sources. He is also the only one to give the dual date for clarity.

"The 20th of January, being Tuesday in the morning the wind blowing hard at south-west, there was so great a flood at high-water, that the sea broke over the banks, and overflowed all the marsh country in England and in Wales, drowning their cattle, and carrying away their corn and hay, some horses and many trees. Some lost their lives, and many saved themselves by climbing up on the roofs of their houses, and others on trees and mows [haystacks]. In the marsh country about Aust and Henbury, the flood was so high it could not all run off again, but remained a fathom (6 foot) deep, and the people on the trees could not come down but remained there 2 or 3 days." This fits with the details in the various woodblock prints, suggesting a shared, but now unknown, source.

"The mayor, Mr Barker, hearing thereof commanded cock-boats [from Cork, Ireland] to be hauled thither to fetch them off, that they might not perish. In the city it rose on the Back 4 ½ feet above the street: so that a small boat about 5 tons came up laden to St Nicholas crowd [crypt] door; and the boatman put his hook against the lower step and thrust off his boat again. All the lower part of the city was covered; it was in every house on the Back and most part of the Key, doing much hurt in cellars to woade [blue dye], sugars and salt: butts of secks [sherry, the city's main import] swam in the cellars above ground therefore worse in vaults under ground. In Redcliffe, Temple and St Thomas streets (in the southern, low-lying region) it was half way up the seats. The Bridge was stopped [blocked] and the water bayed back higher towards Redcliffe-street.

"It rose 5 feet at Trin-mills. At it's return it brought great trees down the river, but did no harm to the bridge."[5]

This shows the flood continued upstream, raising questions as to whether it reached Bath. But as this part of the river was not canalised, more likely its force was dissipated after leaving Bristol.

The Gentleman's Magazine of July 1762 cites a pamphlet in the Harleian Library dated 27 January 1607, apparently the modern date: "The first bursting of the sea over the banks in prodigiously high waves", which he describes as "tremendous. The whole vale from Bristol to Gloucester was overflowed for 6 miles distance from the river on both sides; and most of the bridges and buildings were destroyed."

The parish registers for the large parish of Henbury, which included many small settlements, records that Alice Hurne and her daughter Edith drowned in their beds and were buried on 22 January.[6] As with the victims in Chepstow, this was late for them to be in bed. Had the woman just given birth? A further 6 bodies were later found as the floods receded. The final victims were found 5 weeks after the flood: 6 people from the household of Thomas Longe, giving an indication of how long the clean-up continued. At Almondsbury, the cover of the register lists that a married man, a youth and 2 unnamed children were lost. At Oldbury, 2 men and a woman were lost.[7]

The only other named witness to the disaster was Sir John

Stradling(1563–1637), who was Sheriff of Glamorgan at the time and was rescued from Aust. He was probably an important source of details on the disaster. He claimed, "I have seen fish and men hanging from trees, while the cow, sheep, and horse swim in the sea."[8] This is yet more evidence of how the disaster was seen as the world being turned upside down, a situation which was repeated when James 1 was executed.

Ernest H Baker's 'A True Account ...' provides: "BRISTOW Now bend your eies upon the Citty of Bristow, and here beholde as much cause of lamentation as in any place of this realme, that hath tasted of the like misery. In the self same Moneth of January, and much about ye very day, did an arm of the North seas break in [at a spring tide] which overflowed not onely the banckes, but almost all the whole Country round about."[9]

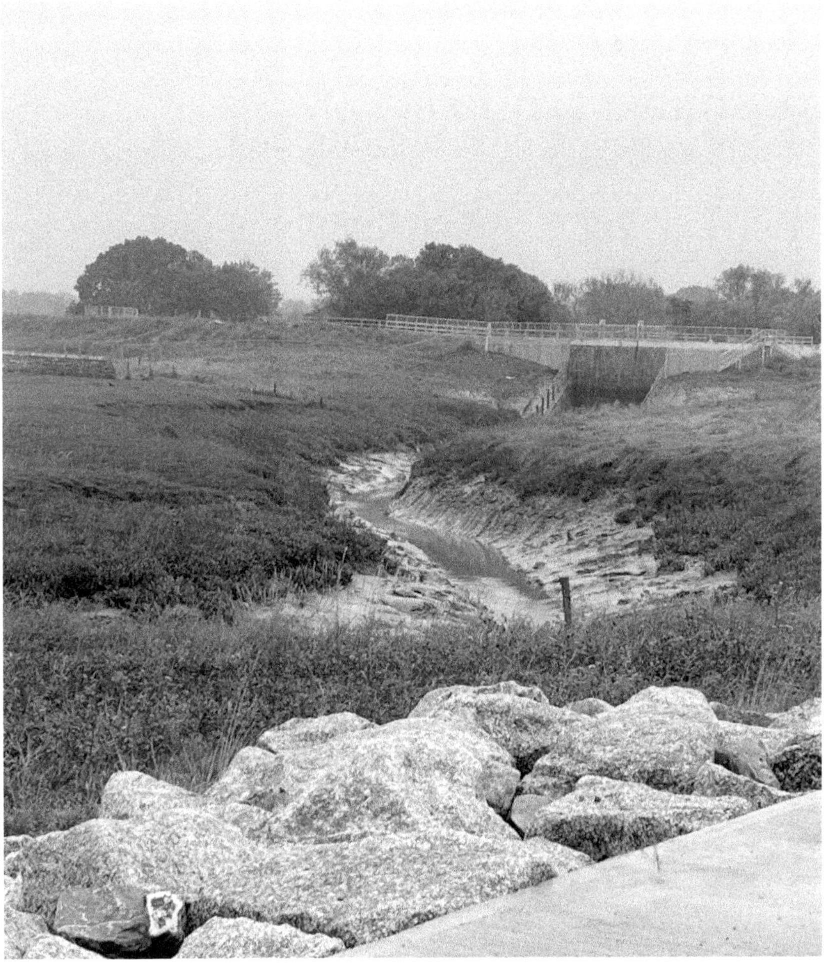

## CHAPTER 15
## THE GREAT FLOOD IN SOMERSET

Somerset's name derives from the Saxon term for 'summer settlement', as the region was uninhabitable in winter due to flooding, when fertile silt was deposited. This meant grass was available earlier than elsewhere, averting the annual dearth at the end of winter and providing lush grasslands for those who brought their beasts from higher ground to feed on the pasture. Thus what are now often considered to be dangerous floods were then merely extreme forms of the natural cycle, essential to feed the region. The main problem now is that human settlement has become permanent and is increasingly struggling to adapt to climate change.

Modern media has given us access to terrifying images of flooding, but such events in Somerset are not new. Following the floods of 1875 the press claimed never before "have the floods attained such a height, covered so enormous an area and caused so much loss and misery".[1] The region's main river is the Parrett which is tidal for 27 miles inland, so when runoff from heavy rains meets high tides, flooding results. Strong onshore winds make conditions worse, and as many rivers are higher than the surrounding land, protective banks need to be constantly maintained, so the situation is always fragile.

For much of the region's history, many locals were poor and lived in houses made of mud and clay which dissolved when floods hit. This probably explains why numerous settlements named as victims of 1606/7 cannot now be traced. Survivors sought safety on higher ground and it seems some never returned. Over the ensuing centuries, many schemes for improvement have been proposed, which mostly involve digging more drains. But funding was always a problem, and as drainage raised the value of property, poor locals feared losing access to their common land. People in urban centres should not have objected to the flooding of fields as they gained no direct benefit from it, but many urban workers had part-time work in agriculture, the reverse of the normal pattern due to the richness of the lands, so it was uniquely profitable.

Following the departure of the Romans, coastal regions increasingly flooded as the skills and organisation of manpower were no longer available to maintain flood defences and water management. On rivers such as the Parrett, water management encouraged water transport, but this often conflicted with fishing where weirs and nets obstructed vessels. Due to the extreme weather from the 13$^{th}$ century, Glastonbury Abbey invested much effort in the Parrett, building and maintaining embankments, sea walls and sluices. But after the Reformation, the loss of relevant skills, resources and cheap manpower must have raised the risks of flooding.

The county of Somerset has the largest area of low-lying land in the UK, with most of it less than 8 metres above sea level, so below the high-water mark of spring tides. Much of it is also very flat, so when it floods, the water is slow to drain away. The many drainage ditches and streams and the rivers Axe, Sheppey, Brue, Carey, Yeo, Tone and Parrett all drain into the Bristol Channel, so there were many sites of entry for the flood waters in 1606/7. To compensate for this, land was claimed by the king, who established a Court of Sewers to maintain waterways, drainage ditches and embankments. King's Sedgemoor Drain was built in 1791–5 to divert the River Cary to discharge into the River Parrett near Bridgwater, which provided some relief to the region, but it also raised the risks of flooding in the town.

The annual floods deposited fertile silt, so grass was available earlier than elsewhere, so averted the annual 'hungry time' at the end of winter. Thus, floods are extreme forms of the natural cycle. The modern problem is that human settlement has abandoned seasonal migrations to form permanent settlements, and so are increasingly struggling to adapt to the added problems of climate change.

In 2014 the levels suffered the worst floods in living memory, with an estimated 10% of land under water. The town of Glastonbury was flooded, despite its nearest beach being Burnham-on-Sea, almost 14 miles away.

Modern media has given us access to terrifying images of flooding, but these events are not new. The region's main river is the Parrett which is tidal for 27 miles inland, so when runoff from heavy rains meets high spring tides, flooding results. Strong onshore winds make conditions worse. Due to centuries of silt deposition, many rivers are higher than the surrounding land, protective banks need to be constantly maintained, so the situation is always fragile.

At the time of the Great Flood, victims sought safety on higher ground and it is possible some never returned. Schemes were proposed to protect people and property and new ones continue. But funding is always a problem. The cost of improvements had to be recouped, often by raising rents, but this did not increase output, so poor locals feared losing access to common land via enclosures.

The Great Flood swept up from the Atlantic, with waters rising and becoming more violent as the Bristol Channel narrowed. The claim that Brean was "devoured" seems extraordinary as the peninsula rises 200 foot above the sea, but this must have referred to the coast rather than the promontory which juts into the channel. Both Flat and Steep Holms are on the same line of rock projecting into the Severn, used for sheep rearing, so the sheep and shepherds were likely swept away.

The River Avon forms Somerset's northern border, so the following list of possible flood victims starts near its mouth at Portbury, dominated by its former marina which is now a housing estate. It is older than Bristol and was an important Roman port when Bristol was a mere river crossing. The Romans possibly built ships there and shipped Mendip tin and lead to their settlements in South Wales. Beyond the M5 to the south is St Mary the Virgin's church, part of the port and settlement founded by the Normans. The coast has a moth-eaten appearance of standing water and waterways for transport and drainage which is common in the region, showing its ongoing risk of flooding.

Portishead is one of the many places to claim the highest tidal range in the world, up to 45 feet. Its headland was fortified by the Romans, and the Saxon Wansdyke ends in a nearby field. It has a lighthouse at Nore Point. It projects into the channel so must have taken much of the flood's force. The church there dates from the 14$^{th}$ or 15$^{th}$ century, so it survived the Great Flood, and still includes a rare musician's gallery in the porch. But the rest was altered and enlarged in the late 19th century, so the storm seems to have left no trace. The nearby court has windows from the 16th century so these seem to be rare survivors, with mostly Victorian buildings towards the coast. Woods survive on the headland and maps show a place called Dry Hill, suggesting the surrounding area was often otherwise. Maps show small rivers and brooks, and a particularly large number of reens or rhynes, built for drainage. The hill is surrounded by 6 others, ranging from 100 to 300 feet where the Severn meets the Bristol Channel, with views to the Mendips and South Wales. It has a huge cemetery along the water's edge, reflecting the numbers of maritime losses, and the deaths of people who visited for their health. When visibility is good both the Holmes are visible. Its headland was fortified by the Romans, and the Saxon Wansdyke ends in a nearby field. It has a lighthouse at Nore Point.

North of Clevedon on the coast is a golf course, a popular use for poor quality land, in this case, exposed to the elements.

Clevedon is described as "gathered about a rocky bay" but is mostly Victorian. Its Norman church is on one of 7 hills, but as late as the 19$^{th}$ century it had a cluster of cottages around it, so there seems to have been little to be flooded.

In the parish of Kenn between Clevedon and Yatton, near Ken Moor, is the small 19$^{th}$ century church of St John Evangelist; as its name suggests, it is situated on marshy ground, so it is at high risk of damage from storms, including that of 1606/7.

Several miles south is Kingston Seymour, the region's best-known victim of the 1606/7 flood. It has a pair of boards in its porch, one of which details the disaster, and has already been cited in E E Baker's book as evidence of the event. Mee claims "The sea has several times tried to swallow this rich marshland village... Painted on wood in the church is an old account of he time when the waves broke down the village defences and filled the Norman font with sea water. For ten days the flood was 5 feet deep".[2]

Missing from all these sources is the second board in the porch: it records a second flood in 1702 which led to a lawsuit between the church and St Mary's Yatton, reflecting the ongoing need for co-operation between parishes to fund defences. But is this the wrong date? Was this the Great Storm of 1703, as recorded by Defoe, or another event? Or is this yet another example of confusion over the calendar system? The list of rectors in the church includes John Seaward who was instituted in 1616.

The church is low-lying and on a large loop of a stream and in the churchyard is an unusual lych gate-bridge combination. Architectural historian Nikolaus Pevsner makes no mention of the board or of ongoing flood risks, confirming how unlikely he was to mention lesser-known sites.

William Camden in his 'Britannia', of 1607,[3] is the only contemporary source to provide a clear timescale for the disaster, which apparently lasted longer than other sources claim: "After a spring tide, being driven back by a south-west wind (which continued 3 days without intermission) and then again repulsed by a very formidable sea, the wind raged with such a tide." This is confusing as it mentions a strong wind and apparently a second high tide. Did the flood last for several days, which explains the huge range and variety of accounts?

To the south west the land is well supplied with rhines for drainage and Sand Bay continues the pattern of poor quality coastal land at risk of floods and erosion. This echoes the situation in South Wales where such land is now mostly holiday accommodation.

Weston-super-Mare is on a wide bay, a fishing village which became a Victorian resort and is now largely a commuter town for Bristol. It was the ancient settlement of Weston-on-the Moor, so poorly drained and at risk of flooding. It was still mostly sand which was driven inland in the early 19th century when the sea wall was built which allowed development. It suffered damage in storms in 1981,

1990 and 1996. Much investment has produced protective sea walls as much of it is below sea level.

Uphill is beyond the southern edge of the Weston, north of where the River Axe joins the Bristol Channel and is sheltered by Brean Down. It was possibly the site of a Roman fort. From the 16th century most of its trade was the importation of lime from South Wales to improve the soil for grass to fatten cattle for Bristol's markets. By the early 20th century this was a major industry. No date is given for the following: "All the swampy moorland around Uphill for many miles, through the medium of dykes, was transformed by them into fertile pasture".[4]

The old church was called the "Beacon on the Hill", and used by mariners navigating the Bristol Channel. Churches on hills are generally dedicated to St Michael, but here its role in helping sailors navigate means it is dedicated to St Nicholas, their patron saint.

Inland is a local nature reserve, showing the land is still of low value, likely due to its risk of flooding. Was the parish still swampy at the time of the Great Flood, and have all the records been lost? On top of the hill is a roofless Norman tower, now cared for by the Churches Conservation Trust. It was probably a place of safety when the flood arrived. Its last service was in 1846. The present church is on lower ground. The 2012 floods on the coast hit Clevedon badly, but especially West Huntspill, and Sand Bay where the waters reached up to 6 metres.

Brean Down is a limestone headland, 318 feet high, part of a ridge that extends to Steep and Flat Holms before erosion left them as isolated outcrops to the north of Bridgwater Bay. To its south and inland is Brent Knoll, the landmark for telling travellers on the M5 that they are approaching Bristol. Flood victims likely sought shelter and relief there. Many historic floods came from the south west, and the Romans probably built some of the embankments, and possibly a fort there. In the reign of Edward III it was used for rabbit warrens, so was poor farmland but a valuable source of fresh meat in winter.

Breane: this is a long sandy stretch north of Burnham-on-Sea. Problems with flooding there were of long standing even in 1483 "Thomas Baret … suffragan bishop to the See of Wells, 1483, which included the lordship of Brean, in Brentmarsh, on condition that he applied the revenue "to fortifie the sea walls and banks for the salvation of the said lordship."[5] The church is dedicated to St Bridget, suggesting it was isolated, poor quality land. Its spire was whitewashed as a landmark for shipping. The Great Flood broke the sea banks there, flooding 30 villages.

It is now mostly a village of caravans, confirming the land is still of poor quality. Wikipedia claims it was an Island before the levels were drained. Known as the Isle of Frogs, it was a safe haven from from the marshes, especially during flooding.

Pevsner mentions the church in passing as being "much renewed", but Mee provides the following: "It has grown into a straggling collection of bungalows with a long narrow green jutting a mile into the sea, but 200 feet above, overgrown with grass and wildflowers, is a sanctuary for wild birds. They have another sanctuary on Steep Holm, the island a little way out to sea, with Flat Holme beyond it. ... For years the north side of the churchyard [of St Bridget's whose tower was struck by lightning] was set apart for sailors drowned on the treacherous Brean Sands when Bristol's ships were constantly passing this way."[6] Baker's booklet of 1884, cited in chapter 7, claims "besides other villages standing in valies is Brian Downe, a village quite consumed". Chilling.

Writing in the early 20$^{th}$ century, Robinson claimed "Years ago there seems to have been much ague (malaria) in Brean, ... that some victims "shook so violently as to cause their houses to shake.""[7]

Burnham-on-Sea: Mee describes the church as follows: "Its tower is leaning in the sinking sands, for its 15$^{th}$ century church stands almost on the shore of this bright little seaside town. There is a lighthouse which flashes a white light to warn ships from Gore Sands."[8] This suggests it was a flood victim, but Pevsner claims the church was repaired and enlarged by the Victorians, so no records of the flood damage survive. He adds: "there are clear traces of the village before it became aware of the sea."[9] This is hard to make sense of but seems to be suggesting the sea has become more intrusive and violent than in the past. It seems urbanisation has caused any remaining traces of the flood to be lost.

The southern boundary of Burnham is the River Brue which reaches Bridgwater Bay near the mouth of the meandering River Parrett after it has been joined by the River Huntspill. The term 'pil' is of Welsh origin and a common term on the Somerset coast, meaning a tidal inlet/harbour. The settlement of Huntspill, named in the Domesday Book on the mouth of the River Brue, was an extensive Roman/Saxon haven which silted up in the medieval period. The parish's west boundary is the tidal River Parrett which changed its course, so some parts are in the parish of Otterhampton. Its church of St Peter dates from 1208 but it was gutted by fire in 1878. It was a victim of the 1606/7 flood.[10]

South of Huntspill is Pawlett, another likely victim of the flood: "stretching out to sea are the Pawlett Hams, 2000 acres of the richest pasture in Somerset, once belonging to John of Gaunt. His own lands lie below sea level with beds of peat sometimes 30 foot deep". But "they are securely banked from the tides and drained by dykes. The village is safe on high ground." The only hint of problems is in the parish church "the line of ancient seats, with doors into them and carved ends, wave up and down where the floor has sunk a little".[11]

The River Parrett was wide enough to have had a ferry at Combwich where the Victorian parish church is safe upon a hill. Its lych gate unusually commemorates the men of the parish who returned from the war, rather than those who died, suggesting they are people used to loss. Downstream the landscape is again a Swiss cheese of water with the c.500-hectare brackish Steart Marshes Wetlands Reserve again recording human abandonment to the tides. Steart Island broke off from the mainland about 1798, meaning the coast extended further into the sea in 1606/7. The Steart Marshes became the Wildfowl and Wetlands Trust's (WWT) first working wetlands to protect the mainland from extreme weather. Steart church is claimed to be the "only dry place on the River Brue to its north".

Stockland Bristol parish was often flooded, so its church needed rebuilding in 1865. The worst flood was in 1981. The 15th-century font and restored medieval screen survive in the church, showing it was a significant settlement, and its location suggests it was another victim of 1606/7. It is owned by Bristol Corporation, and is on Combwich Reach where the River Parrett reaches the sea, and which was often flooded in the past millennium.

From Roman times there was an important ferry across the river here, and local produce was exported until it silted up in the 1930s.

Cannington is 5km north west of Bridgwater on the west bank of the River Parrett, with a brook running through it, so must have flooded. The Victoria County History describes it as a large ancient parish where the Saxon "Herpath" entered the parish via the crossing. It sits on the flats of the Parrett estuary where Pevsner claims "the character of the landscape here is not Somerset, but rather East Anglia or even Dutch". Leland called it a "pretty, uplandish town", which by the 1840s was genteel with salubrious air. It included the ferry crossing of Combwich, another name with echoes of Wales, suggesting it is of ancient origin. To the north are marshes which drain into the River Parrett. It has suffered coastal erosion, which has been limited by modern flood prevention schemes, so it likely suffered in the Great Flood.

Otterhampton is described by Mee as "a tiny place near the mud flats of the River Parrett."[12] So another likely victim of the Great Flood.

Weston Zoyland is far inland, west of Bridgwater but in the winter of 1630, its commons were under water. The following spring, poverty was widespread so to preserve food, local authorities restricted the work of maltsters. In 1673 and 1678, winter floods forced people to travel about by boat. Large areas, especially south of the village were classed as moor in 1638, probably due to neglect of the drainage, which continued to be a problem into the 1780s when the region was under water for several months each year. The problem here seems to be caused by extreme weather, but it mostly reflected the expansion of agriculture in regions where traditional land use was being replaced by drainage, and 'improvements'.

Bridgwater was Somerset's second port after Bristol, and is listed in several early records of the flood. But its history is dominated by the history of Robert Blake. It is on the River Parrett which is famous for its bore. No record can be found of it being flooded in 1606/7, though it is surrounded by drains and rhines, so it must have suffered.

Watchet is 15 miles west of Bridgwater at the mouth of the River

Washford on Bridgwater Bay. From the 10th century its trade increased despite it being damaged by severe storms. From 1564 the merchants were importing salt and wine from France. The parish church is high above the harbour so would have provided a refuge for victims of the great flood. Its "Primitive jetty" was damaged by a storm in 1659 so was rebuilt in 1708 as a stronger pier. It was damaged by the 1859 Royal Charter storm.

Minehead was originally called 'mynydd', Welsh for market, reflecting the trans-Severn trade. It is described by Mee as 'the sea gate to Exmoor', comprised of 3 parts. On the steep hill that rises 850 feet above the sea was the old Church of St Michael, which was a landmark, providing a guiding light for mariners from its rood loft window. The fishing village was by the quay, and from 1800 the seaside resort grew alongside. Carved above the east window on the outside wall is an old prayer for the town's fishermen: "We pray to Jesu and Mary, Send our neighbours safety".[13]

But Wikipedia claims it lost its town charter in 1604 as the harbour had silted up and the port fell into decay. A new harbour was built at a cost of £5,000 further out to sea, which included a pier in 1616. It is thus unclear how badly the town suffered, but this seems a long time to take to rebuild a harbour, suggesting the storm interrupted the engineering, and raising funds took time in its aftermath.

Its history is well recorded, so it seems odd that no mention can be found of the Great Flood, as its port must have suffered, with victims seeking shelter on higher ground, especially in the church. The town was described by Defoe as Somerset's best port and safest harbour. It had extensive trade with Ireland, and later with Virginia and the West Indies. The town's sea defences were upgraded after the 1990 storm, so it is still at risk.

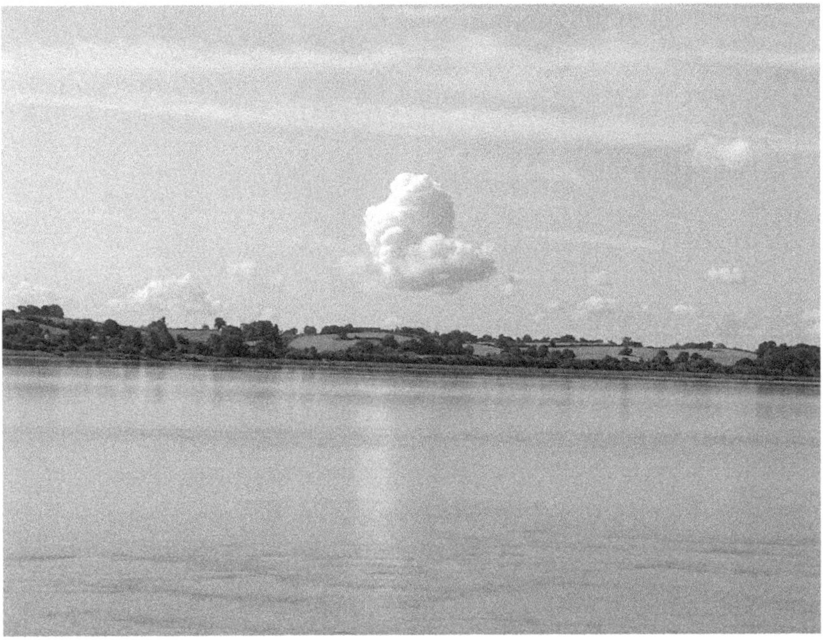

# CHAPTER 16
# BARNSTAPLE AND BEYOND

Baker's account claims that parts of eastern England were victims of the Great Flood, which means it must have damaged coastal regions along the South Coast en route. This means that records from Devon should be able to establish the extent of the Great Flood, as it has 2 coastlines, facing 2 Channels, i.e. Bristol to the north, and the English to the south, so was exposed to and suffered badly from storms sweeping in across the Atlantic. It could be expected to provide records of the Great Flood. If affected sites were only on the north, this would suggest the flood swept up the Severn and the Bristol Channel only, so discounting any claims that it affected the capital and/or the South East. Devon's biggest towns were Barnstaple, which was flooded, and Tiverton. There were 2 upland moors which must have flooded if they were in the path of a storm, with the runoff affecting coastal settlements, but not if it was a flood only. The South West is famous for its Jurassic coast, with dinosaur skeletons discovered after the collapse of cliff faces, exposing the long-buried bones. Most settlements are thus high above the sea, which likely protected them from the floods, including that of 1606/7, but their shipping would have been lost, and the flood is likely to have undermined the cliffs, causing land and buildings to be lost, together with their inhabitants and records.

But unfortunately it is beyond the range of my travels and online history is being swamped with tourist information for this popular summer holiday destination. Mortehoe, near Woolacombe is described as "a quiet peaceful village with dramatic scenery and abundant sand". It was, like much of the coast, famous for wreckers and smugglers, with spectacular views from its cliffs. Mee describes it as "a rugged place, with the solid rock sticking up through the steep streets".[1] Ominously, it is a short walk from Morte (Death) Point, with views of the pirate and smuggler paradise of Lundy Island, all of which suggests the Great Flood could have left records, but they have been lost.

Its church is the Grade 1 listed St Mary, so of ancient foundation. Intriguingly, none of its parish records survive, but the bishops transcripts, the official copies, start in 1607, i.e. very soon after the flood, which suggests the church and its contents were lost.

This may seem like a detour, but the Norman church at Northleigh had a new rector, John Carpenter, a year before the Armada who wrote religious tracts. One of these was 'A Sorrowful Song for Sinful Souls composed upon the Strange and Wonderful Shaking of the Earth, April 6, 1586', so within living memory of those who suffered the Great Flood. It is now called the Dover Straits Earthquake, the best-known disaster of its age as it was widely reported in pamphlets. It struck the South of England — especially London — between 5 and 6pm, though it seems the North and West were spared. It damaged steeples and chimneys, and made some church bells ring, including St Paul's. Its pinnacle, the roof of Christ Church Hospital and Paris's Notre Dam were damaged. Incredibly, only 2 fatalities were recorded. Puritans blamed it on theatres.

The response to it echoes that to the Great Flood, with the event being seen as a sign of God's anger with his people, and a warning to them to mend their ways, or risk the end of the world.

'History Today' claims the Crown and the state Church capitalised on the event by encouraging religious conformity and publishing a pamphlet for guidance in prayers in the national Church. The appropriately named Thomas Churchyard published 'A warning for the wise, a feare to the fond, a bridle to the lewde, and a glasse to the good ... for the glorie of God, and benefite of men that can warely can walke and wisely can judge." Churchyard mocked those who sought natural causes for the event, i.e. early scientists, but compared it with other warnings, such as plague, war, comets and monstrous births.[2]

Lyme (Regius), on the South Coast, is prone to landslides described in the late 19th century as "in every direction one sees garden walls shored up and houses dropping from the perpendicular. Old inhabitants remember when the cliff between Lyme and Charmouth extended half a mile further out to sea than now; and all along the shore rock falls are perpetually taking place so fast that quarrying is not allowed, nor is it necessary."[3] This last phrase has a chilling quality to it.

The source continues: "This frequency of landslips has determined the character of the whole coast for several miles west of Lyme, and has imparted to it a wild and surprising beauty, casting it into the form of terraces and undercliffs, the debris of successive falls which at one time, when the scars were recent, appeared as defacements on the grassy slopes, but in a climate so mild and moist has this have been quickly converted into hanging gardens. A path winding pleasantly through overarching trees leads by a steep ascent up the first slopes above the town, emerging on a plateau, whence one looks down on the sea washing gently among the shattered boulders at the base of a grassy cliff. There are ferns in all the crevices, and trailing ivy on the rocks. The grass is spangled with violets and there and there an orchid glows deep and red in the increasing sunlight. ... in looking across any one of these strangely broken slopes one can easily fancy the surface to be still heaving and subsiding as it did once long ago when the great mass slipped away ... In 1839 some cottagers living in the edges of the cliff were alarmed to find some cracks in their gardens and nearby cliffs which increased as the year wore on, till on Xmas Eve residents returned home to find their garden had dropped by a foot; by the morning it sank 6 feet more. The following night a storm blew up and by morning ¾ mile of cultivated land had fallen, taking 45 acres of arable land, 2 cottages and an orchard, leaving dramatic broken gullies and inspiring a great sense of fear to see the formation of a reef 40 foot high a mile long close to the shore. This earthquake at Axmouth was described as being worse than the earthquakes of Calabria which attracted so many English tourists. The reef was washed away, but the cliff has been covered in luxurious vegetation, inspired tourists who were relived to climb the steep paths and reach solid ground."[4]

The coast is interrupted by long thin valleys, or combes. Camden wrote an edition of 'Britannia' in 1607 which mentioned the collapse of chalk cliffs, but it seems not of the Great Flood. Leland in the reign of Henry VIII wrote of a "notable haven" at Seaton, but "now ther lyith between the 2 points of the old haven a mighty ligge and barre of pible stones in the very mouth of it, and the River Axe is driven to the very east of the haven called Whitclif and there at a small gut (gout/drain) get into the sea and there come in small fishar boats for soccar".[5] This suggests the port had declined due to its failure to protect itself, through idleness or lack of skills formerly supplied by the monasteries. The region had made little contribution to the nation since it sent 2 ships to Calais in 1343.

Between Seaton and Beer is a crumbling white cliff. Smuggling was common till the late 19th century.

Teignmouth lacks a deep harbour, and its entrance is through sand-banks, so it was never suited to large ships. Sheltered from east and west winds, with a bay almost emptied twice daily, as was the case in Bristol before its Floating Harbour was built, this was a problem for large ships, and it was largely excluded from many of the great events in history. The banks may have protected it from the Great Flood if it reached there, with the bay and hinterlands dissipating much of its force.

Shipping was mostly confined to bays defended by chains across the entrance, eg Torbay, which was chosen for the arrival of William III.

There is a legend of a lost land beneath the sea between off Lands End, i.e. the location of the Isles of Scilly. It was a large stretch of fertile country of 140 parish churches comprising the Lost Land of Lyonesse, with the Isles of Scilly surviving as former high points. Remnants of ancient forests are still found beyond Cornwall, including moults of beech trees with nuts at the mouth of the bay, suggesting the inundation there was sudden, and in autumn, which fits with the stories, confirmed by archaeology off the coast of Wales.

The 'Anglo Saxon Chronicle' of 1014 records on St Michael's Eve "a mickle sea flood [came] widely through this land; and it run up so far as never before; and it drowned many towns and mankind too innumerable to be computed."[6] In 1099 the chronicle records "This year sprang up so much of the sea flood and so mickle harm did as no man minded that it ever before did".[7]

There is a charter from the time of Henry I which granted to the monks of Tavistock "all the churches of Sullye with their appurtenances and the land as ever the monks or the hermits in a better state held it during the time of Edward the King." Hermits often gave shelter to victims of extreme weather on dangerous coasts. This suggests some tragedy had caused the Scillies to decline, 15 years after the second and worst of the storms had struck. Stories of lost lands are the stuff of legends but charters and archaeology are not. The above suggests a great tragedy had struck, which has left little trace, so it has parallels with the 1606/7 flood.

It is also worth putting these tales into the context of the similarity in names seen in South Wales and the West Country, as a small island west of Penarth, only accessible at low tide is also called Sully.

Blackmore describes Tiverton in the 19th century where "The schoolhouse stands beside a stream, not very large called 'Lowman', which flows into the broad river of Exe about a mile below. This Lowman stream is wont to flood into a mighty head of waters when the storms of rain provoke it; and most of all when its little mate, called the Taunton Brook … comes foaming down like a great roan horse, and rears at the leap of the hedgerows. Then are the grey stone walls of Blundell on every side encompassed, the vale is spread over with looping waters and it is a hard thing for the day boys to get home to their suppers … and in the very front of the gate, just without the archway … you may see in copy-book letters done a great 'PB' of white pebbles. Now it is the custom and law that when the invading waters, either fluxing along the wall from below the road bridge, or pouring sharply across the meadows from a cut called 'Owen's Ditch'… upon the very instant when the waxing element lips through it be but a single pebble of the founder's letters, it is in the licence of any boy, soever small and "undoctrined", to rush into the great schoolrooms, where a score of masters sit heavily, and scream at the top of his voice

"P.B.". Then with a yell the boys leap up or break away from their standing; they toss their caps to the black beamed roof, and haply the very books after them; and the great boys vex no more the small ones, and the small boys stick up to the great ones; one with another hard they go to see the gain of the waters... Then the masters look at one another ... with a spirited bank they close their books, and make invitation the one to the other for pipes and foreign cordials, recommending the chance of the time, and the comfort away from cold water."[8]

Paignton has a long open beach and produces the finest china clay in Devon. Wikipedia claims its beach once led to low sand dunes with marshes separating them from inland hills, but it declined after the Reformation. To the east is Fairy Cove, described as a former desert.

Appledore is at the mouth of the River Torridge, near the Taw Estuary, an ideal site for trade but at high risk of storms and floods, about 3 miles west of Barnstaple. Its port thrived in the Elizabethan period, but a 60-ton ship was about to sail when the flood hit, forcing it beyond the high-tide level.

Barnstaple: it is hard to find records of the Great Storm beyond this place, which is mentioned in various sources. This is largely due to the wealth of local information in the city's chronicle between 1586 and 1611, which describes the crucial year and provides useful context. It provides the most detailed account of the storm outside Bristol, thanks to this chronicle by Adam Wyatt. The River Taw is to its west with its marshes, the Coney to the south and Yeo to the north, which meant the town was well supplied with local food. The Taw meets the Torridge just before it reaches the huge stretch of sand, the Braunton Burrows, which protect the port from storms. But as ships sailed further, to the Americas, Africa and the Far East, and brought coal from Wales, they became larger, so that from the 1580s the protection became a hazard for shipping. The Burrows are now a vast system of sand dunes, the centre of an area of outstanding natural beauty, protected by UNESCO. Much of it is below the water table and is a former wetland which is still popular with birds and other wildlife. It must have suffered in the storm, but no trace has been left.

Wyatt's chronicle contains unique and valuable details of the time including weather, trade, food supply and maintenance of the town, including its flood defences. The town is especially prone to floods as it is called the 'Town of Bridges'. From 1603, improvements were made to the main quay. Victims of drownings are buried in surrounding churchyards.

The Great Flood of 1606/7 produced a wave 5–6 feet above the highest in living memory. Merchants claimed to have lost £1,000 in goods and a roof fell on a man and his children in bed. The waters forced open locked doors, destroyed wells and houses including a roof which collapsed and killed James Frost and his 2 children. The parish register claims it arrived at 3am and lasted 9 hours, the only source to provide such detail. But this is odd as the often-cited accounts from South Wales claim the flood was said to have arrived at 9am. Baker claims there was another flood the following year. If so, are we here dealing with 2 different events?

The winter of 1607, i.e. either the same year as the Great Flood or the one following, was very cold. Grain was imported from Gdansk to make up the shortfall and big speculators sent prices soaring.

The chronicle also provides valuable context to the flood, some of which was quoted by the Venetian ambassador, so many details became widely known throughout Europe. Late 17th-century Devon suffered repeated harvest failures in 1587, 1588 and especially 1589. There seems to have been food shortages in 1566/7, 1591 and 1595–7 when grain prices soared, and local government struggled to import enough grain to avert mass starvation. Rainfall was high in autumn 1588, sending prices soaring. Nationally, bad harvests are recorded for 1580, 1594–7 and 1608, so 1 or 2 years after the Great Flood. Droughts are recorded for 1587 and 1590, which caused shortages, and further droughts in 1588 and especially 1597.

Gray mentions a tempest arriving in Barnstaple from the west on Michaelmas Eve, 28 September 1586, when the marshes flooded and "ripped diverse houses" including the fish shambles (market) till about 4pm.[9] Another unique claim is that an earthquake hit early on 19 May 1607, apparently 4 months after the Great Flood, though no damage was recorded.[10]

Respite came in 1599 with a bumper harvest, but shortages

returned in 1608. The dearth was aggravated by competition at sea from overseas fishermen, suggesting there were also food shortages in mainland Europe. Grain was being converted to bread and to brewing instead of being exported. Plague struck Barnstaple with over 40 dead in 1579/80, then returned in 1597 and 1604 in South Wales. On 23 December 1599, a violent tempest of wind was recorded by the Vicar of Fremington and several harsh winters followed, especially 1593 when the River Taw froze. Another journal, by Richard Wood, Vicar of Frayne 1533–1678, lists further problems in the region. On 6 January 1604 was a flood, and on 7 December a great frost, but there was a gap till January 1608 when a new walking place was made at Barnstaple, or was this a flood repair?

In 1607/8 the weather was much colder than usual — and Barnstaple flooded and several roofs were blown off — allowing people to walk on the frozen river, as was possible on the Thames in London. Colonists in North America suffered badly from the cold and hunger, which forced many of them to return home.

In 1609 the churchyard was "new righted". Again, was this after the Great Flood, or due to repeated damage over the years or a surge in burials causing overcrowding? The Victoria County History for Tiverton in 1623 records 53 houses were destroyed by floods.

But there is some confusion over these events, as another source claims "a tremendous flood struck in the autumn of 1606". It flowed up the lower end of Cock Street and along Maiden Street "where it destroyed walls and a house, killing the inhabitants". The water was deep enough to cover the merchants' stone on the quay, which suggests it was a table for settling debts and agreeing sales like the famous nails in Bristol. The source claims some work had been in progress from the previous year to narrow and deepen the channel which irritated the Vice Admiral of Devon, claiming it was an annoyance to boats and barges.[11] It seems the work made the channel so deep that horses could not pass. Was this work to allow larger, transatlantic ships to use the port, contributing to the town's floods?

Reverend Seyer of Bristol again provides some clarification, as at the time of the storm, Barnstaple was struggling with its sea defences. In 1567, workers made a new cut in the river there, "by reason of a great compass or fetch about of the water of the said river. The sea banks or walls of the said river upon the north-east part thereof nigh to a tenement in the tenure of R- Popham, were so decayed and worn (not withstanding yearly reparations done, to no small charges) that if the sea should have broken over whereof the inhabitants of the country there nigh to the same, were in great fear, it would have drowned about 10,000 acres of ground beside other great harms which might have ensued thereof. It was therefore prevented and foreseen by the Commissioners of Sewars (sic) which included Sir Morreys Berkeley...with the advice of the best heads of good yeomen of the country, that a new cut should be made straight over."[12] Seyer makes no further reference to Bridgwater until the great storm of 1703.

Tiverton seems to have derived its name from having 2 fords, so it was another site well supplied with water. Wikipedia records it suffering 2 fires. In 1596 most of the town was destroyed, and in August 1612 most of it was consumed, allegedly from a dog fight, so it seems it must have flooded but the records have been lost. The Victoria County History records that in 1623, 53 houses were destroyed by floods. So God's people were not only suffering from floods, but many other disasters at the time.

Did this long period of extreme weather cause people to struggle with, or to lose, their faith in god, helping to fuel the rise in the various forms of Nonconformity, local superstitions and witchcraft?

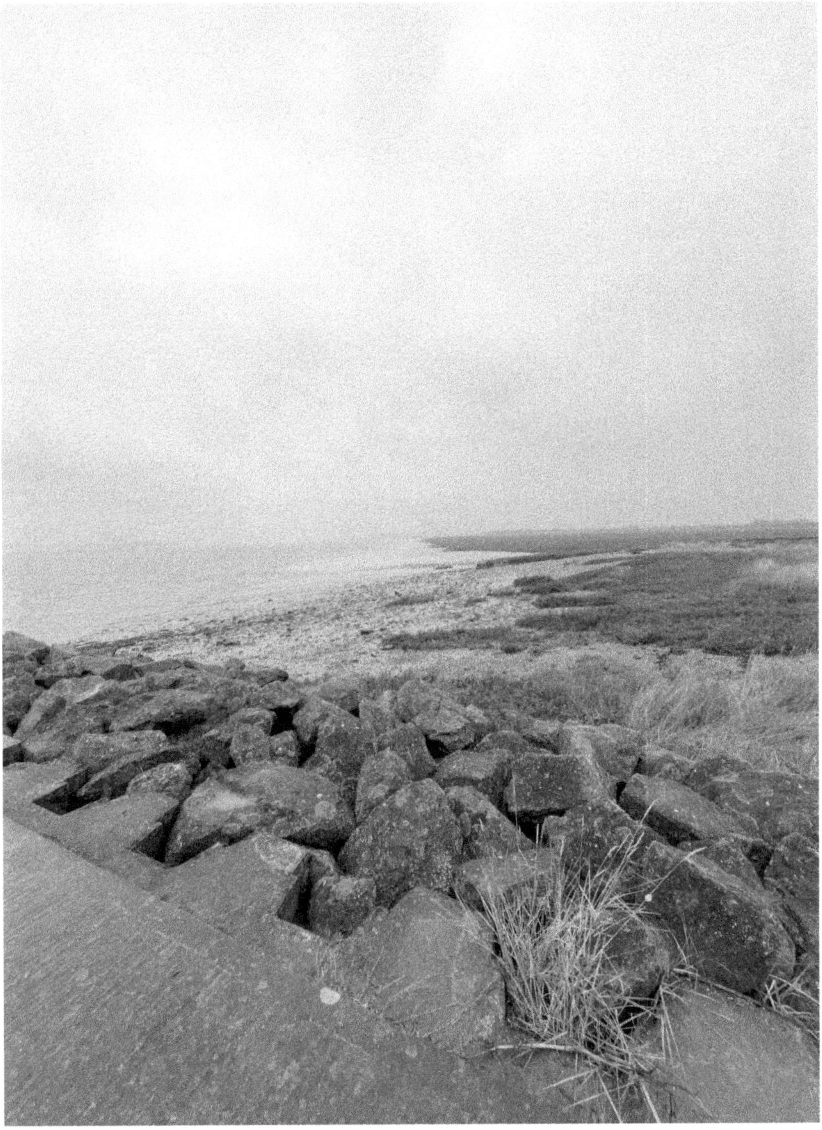

# CHAPTER 17
# CONCLUSION

Thus the Great Flood was not an isolated event in the West of England and Wales but arrived in the midst of a period of extraordinarily turbulent weather which struck both sides of the Atlantic and in Europe. But it wasn't only about weather.

The physical and economic devastation is beyond our comprehension. But in our modern world, this is made worse by our inability to comprehend how our ancestors coped with their belief that they were being punished as they struggled to survive the series of disasters. As with the Black Death, how could they adjust their mindsets, their world view that their loving God could punish them in such horrific ways, and for so long?

The period this book covers was one of the most turbulent of these islands' history. The Great Flood and other disasters in the surrounding decades must have weakened the faith of the victims. It must have made them question their leaders, and even the very notion of a loving God who cared for them. It seems possible that people lost faith in their country in the wake of this horrific weather, but especially the Great Flood of 1606/7. Perhaps the fact that the flood has been largely forgotten reflects that the survivors were unable to comprehend it, to describe or to explain it, causing it to be lost from the historic record.

Britain is well known for founding colonies across the Atlantic, but compared with other European maritime nations, it was surprisingly late to do so.

In my research on early colonialism, I found an intriguing quote from "a contemporary Englishman" who claimed "We are knowne too well to the worlde to love the smoake of our own chimneyes so well that hopes of great advantages are not likely to draw many of us away".[1]

This reflects the English people's famous love of their countryside, present in the many agricultural elements of Anglican worship, especially the celebration of successful harvests. The historian Michael Wood described Shakespeare as a country boy for his love of the countryside. But even in London, people could visit the countryside on Sundays well into the 18$^{th}$ century. The natural world and its landscape inspired the Romantic poets and diarists such as Reverend Kilvert. Before much of the countryside was enclosed it provided free food, building materials and recreation for local people.

John Donne famously preached a sermon in London's St Paul's Cathedral which was later published, promoting the Virginia Colony in the name of God and the King, for which he was granted stock in the Virginia Company. But despite claims that the region was blessed with a mild climate, abundant food and kindly indigenous people, there was little initial interest in the scheme. Why would anyone choose to sail across a vast ocean, in a small boat, leaving friends and family on the basis of mere promises? Something must have changed. That something must have been big.

The founding of the first English colony in Jamestown coincided with a period of extreme weather, especially the Great Flood of 1606/7. Shakespeare's play 'The Tempest' was dedicated to Admiral Sir George Somers (1554–1610), founder of Bermuda, England's first Atlantic colony. He played a major role in rescuing Jamestown by providing it with fresh food and supplies. Were these events linked, or was this synchronicity?

And yet despite the disaster of Jamestown's 'starving time', attacks by indigenous people and the huge expense and risks of sailing in tiny ships, by 1634 the English had settled 8 shires. The population was more than 7,500, including several hundred Africans.

An account written soon after the Great Flood described the affected area as "this beautiful kingdom", with "fruitful valleys" which contrasted with the "great barrenness and famine to ensue after it". In the absence of modern emergency relief, and the dearth of surviving records, how many people lost faith in their land and in their God who had punished them? How many survivors saw the event as a warning of more to come?

The general view is that the English were drawn across the Atlantic by promises of wealth, and that clearly motivated the large investors. They were still looking to the Far East, as early instructions to the Royal Adventurers urged them to settle along rivers, especially those leading to the north west as being the most likely to find the other sea. A contemporary map showed it to be a mere 10 days' march from the head of the James River.

What changed to drive so many people to seek new lives on the other side of the Atlantic? What if they were driven by the extreme weather, and attracted by promises of settling in a new Eden across the Atlantic? What if the Great Flood provided the tipping point that convinced people of the West country and South Wales to fall out of love with their own hearths?

George Thorpe (1572–1622) of Berkeley was involved in founding the Berkeley Company and in the Gloucestershire Lists of able bodied men fit for military service was listed as an Esquire, with 7 personal servants, so was a man of significant wealth who lived in the hamlet of Halmor [salt marsh?] in Berkeley County.[2] In 1609, 2 years after the Great Flood, he leased part of the New Grounds along the Severn shore at Frampton and Slimbridge for 51 years. But in late 1620 he sailed to Virginia. In the interim he and several others had been involved in litigation over the ownership of the lucrative fishing rights which his ancestors had held for many generations. As mentioned previously, this region had been the subject of disputes for centuries, with Slimbridge and Frampton's grounds expanding by 200 acres of very fertile land, called the Warth, at the cost of Awre. Gethyn-Jones claims the drawn-out disputes were incredibly expensive, and "were one of the two main reasons for the sale of much of his estate before he left England in 1620".[3] The other reason was the cost of financing the Berkeley Company.

It seems Thorpe had given up hope in making acceptable profits from the Severnside lands, so sought a new life across the Atlantic. Did the Great Flood of 1606/7 and the many other extreme weather events around this time turn Thorpe away from his homeland? And if so, how many less fortunate local people, struggling to cope with the extreme weather and the constantly shifting landscapes, also cause them to lose the deep fondness for the smell of their own hearths? The common cause of emigration at the time is often said to have been poverty. But what if the poverty was due to the constant procession of storms and floods, especially the biggest of all, in 1606/7 ? The question may have shifted to become, why would anyone wish to stay?

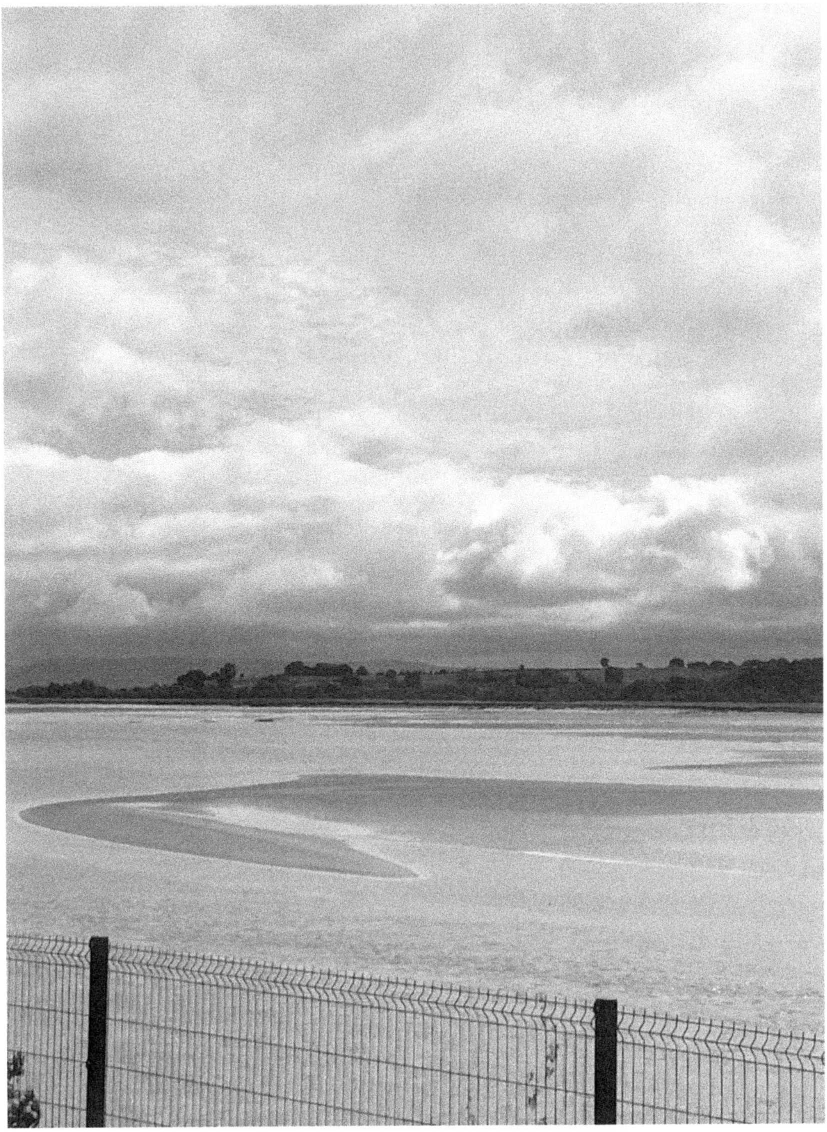

# BIBLIOGRAPHY

**Introduction**
1 139/40 Hutton, E., Highways and Byways in Gloucestershire, Macmillan and Co., London, 1932

**Chapter 1**
1 49 Smith, B., & Ralph, E., A History of Bristol and Gloucestershire, Darwen Finlayson Ltd, Beaconsfield, 1972,
2 142 Bradley, A G., Highways & Byways of South Wales, Macmillan & Co., London, 1914
3 199/20 ditto
4 58/9 North, F.J., The Evolution of the Bristol channel with Special Reference to the Coast of South Wales, National Museum of Wales, Cardiff, 1955
5 59 North ditto
6 61 North ditto
7 112, Rees, V., South-West Wales, A Shell Guide, Faber & Faber, 3 Queen Square, London
8 58 North,
9 58 North
10 76 Rees
11 179 Rees
12 109 Rees
13 52 Swansea Museum Services, Swansea Before Industry, Vol. 1, The Town and Its Surroundings, 1998
14 52 ditto
15 55 ditto
16 55 ditto
17 77 North
18 127 Rhys
19 95 Rhys

**Chapter 2 Records on the Ground**
1 98 Walker, M. Church Life in Sixteenth-Century Swansea, Jones, O W., Walker, D., Links with the Past: Swansea & Brecon Historical Essays, Christopher Davies, Llandybie, 1974
2 109 Walker
3 98 Walker
4 81 van Dam, P J, An Amphibious Culture: Coping with Floods in the Netherlands
5 26, Waters, I., the Port of Chepstow, the Chepstow Society, Chepstow, 1977
6 193 Robinson, W.J., West Country Churches volume 3, Bristol Times & Mirror, Bristol, 1913
7 73 Robinson, WJ., West Country Churches vol. 4, Bristol Times & Mirror, Bristol, 1916
8 2 Robinson, WJ., West Country Churches vol. 3
9 Robinson, WJ., West Country Churches vol. 4,

10 80 ditto
11 163 Barber, C., ed, Hando's Gwent, Blorenge Books, Abergavenny, 1987
12 155 ditto
13 157 ditto
14 137 Robinson, W.J., West Country Churches volume 2, Bristol Times & Mirror, Bristol, 1914
15 178, Long, E.T, ed, Mee, A., The King's England, Gloucestershire, Hodder & Staughton, London, 1966
16 137 Robinson, W.J., West County Churches, vol 2
17 105 ditto
18 xxxix Hewlett, R., ed, The Gloucestershire Court of Sewers 1583-1642, Gloucestershire Record Series vol. 35, The Bristol and Gloucestershire Archaeological Society, 2021

**Chapter 3 God's Warning to His People of England**

1 xi Hamblyn, R., ed., Defoe, D., The Storm, Penguin Books, London, 2005
2 98 Walker, Church Life in Sixteenth-Century Swansea, Jones, O W., Walker, D., Links with the Past: Swansea & Brecon Historical Essays, Christopher Davies, Llandybie, 1974
3 56-7 Notes on the Great Flood of January 20th, 1607, journals.library.wales

**Chapter 4 Lamentable News From Wales**

1 74 North, F.J., The Evolution of the Bristol Channel with Special Reference to the Coast of South Wales, National Museum of Wales, Cardiff, 1955
2 75 North
3 266 Davies, A History of Wales, Penguin, 1994, London
4 xxxix Hewlett, R., Ed., The Gloucestershire Court of Sewers 1583-1642.
5 29 Hewlett, R.,
6 380 Pevsner, N., Founding Editor, Newman, J., The Buildings of Wales, Gwent/Monmouthshire, Yale University Press, New Haven, 2002
7 153 Pevsner
8 263 Pevsner
9 417 Pevsner
10 469 Pevsner
11 www.medievalheritage.eu
12 484 Pevsner
13 511 Pevsner

**Chapter 5 Seyer's Memoirs**

1 259 Seyer, S., Memoirs Historical and Topical of Bristol and Its Neighbourhood, From the Earliest Period Down to the Present Time, Vol. II, J.M. Gutch, Bristol, 1823
2 260 Seyer
3 do
4 do
5 260/1
6 do
7 261 do
8 262 do
9 262 do

10 34, Latimer, J., The Annals of Bristol in the Seventeenth Century, William George's Sons, Bristol, 1900
11 290 Seyer
12 558 do

**Chapter 6 Baker's Reprint**
1 39 Mee, A., revised Long, ET, Somerset, The King's England, Hodder & Stoughton, London, 1968
2 40 Mee ditto
3 53, Robinson, W.J., West Country Churches volume 1
4 32, Latimer, J., The Annals of Bristol in the Seventeenth Century
5 240 Seyer, Rev S.
6 56 Lamplugh, L., Barnstaple, Town on the Taw, Phillimore, Chichester, 1983
7 344, Nicholls, J.F., & Taylor, J., Bristol Past and Present, Vol. III Civil and Modern History, J W Arrowsmith, Bristol, 1882

**8 The Great Flood Tracked to Cardiff**
1 219 Rhys, E., The South Wales Coast, From Chepstow to Aberystwyth, The County Coast Series, T. Fisher Unwin, London, 1911
2 219-20 ditto
3 197 Bradley, A.G., Highways & Byways of South Wales, Macmillan & Co., London, 1914
4 218 Rhys
5 199 Rhys
6 17 Gower Society, The Churches and Chapels of Gower, 2007
7 swanseastmary.co.uk
8 52 Swansea Before Industry: The Town ad Its Surroundings, Swansea Museum Service, vol. 1, 1998
9 126 Rhys
10 128 Rhys
11 130 Rhys
12 131 Rhys
13 133 Rhys
14 146 Rhys
15 252/3 Evans, CJO, Glamorgan: Its History & Topography, Cardiff, William Lewis, 1945
16 371 Evans

**9 The Flood Tracked to Gwent & Newport Levels**
1 141 Barber, C., ed., Hando's Gwent, Blorenge Books, Abergavenny, 1987
2 134 ditto
3 136 ditto
4 522 Newman, J., The Buildings of Wales, Gwent/Monmouthshire, Yale University Press, London, 2000
5 129 Barber
6 22/3 Robinson, W J., West Country Churches, vol 2, Bristol Times and Mirror, Bristol 1914
7 23 Hando, F., Out And About in Monmouthshire, R H Johns, Newport, 1964
8 233 Newman

9 143 Barber
10 143 Barber
11 63 Lewis, J M, Welsh Monumental Brasses: A Guide, National Museum of Wales, Cardiff, 1974
12 379 Hutton, E., Highways and Byways in Gloucestershire, Macmillan and Co., London, 1932
13 146 Barber, Ed, Handos Gwent, vol 1
14 146 ditto
15 156 Graves, C., Tir: The Story of the Welsh Landscape, Calon, Cardiff, Wales
16 373 Newman
17 151 Barber, vol.1
18 151 Barber
19 154 Barber
20 155 Barber

**10 Why Not Wye?**
1 15/16 Clammer, C., & Underwood, K., The Churches and Chapels of the Parish of Tidenham, Their History and Architecture, Tidenham Historical Group, Chepstow, 2014
2 39/40 Mee, A., Ed, Monmouthshire: A Green and Smiling Land, The King's England, Hodder & Stoughton, London, 1951
3 3 Sherborne, J.W., The Port of Bristol in the Middle Ages, Bristol Historical Society, 1971
4 37 Mee
5 15/16 Rhys, E., The South Wales Coast From Chepstow to Aberystwyth, The County Coast Series, T. Fisher Unwin, London, 1911
6 fl_1607_bristol_channel_floods
7 42 Peterken, G., Wye Valley, Collins, London 2008
8 90, Kissack, K., The River Wye, Terence Dalton Ltd, Lavenham, 1978
9 35 Kissack
10 71 Snelling, R., ed., Ordnance Survey Leisure Guide, Forest of Dean and Wye Valley, Publishing Division of the Automobile Association, 1993
11 87 Waters, B., Severn Tide, Dent, J M., London, 1947
12 240 Peterson
13 56 Kissack
14 64 ditto
15 84/5 ditto

**11 Severn Estuary Tidelands**
1 6 C., Witts, The Mighty Severn Bore, River Severn Publications, Gloucester, 1996
2 15 Smith, B., & Ralph, E., A History of Bristol and Gloucestershire, Darwen Finlayson Ltd, Beaconsfield, 1972
3 314 Hampton, Lord & Pakington, R., eds, Mee, A., The King's England : Worcestershire, Hodder and Stoughton, London, 1968
4 14 Long, E T, ed, Mee, A., The King's England, Gloucestershire, Hodder and Stoughton, London, 1966
5 15 Smith, B., & Ralph, E., A History of Bristol and Gloucestershire, Darwen Finlayson Ltd, Beaconsfield, 1972
6 85 Bowers, D G., The Tides of Wales – The Compact Wales, Llygad Gwalch Cyf, 2021

7 24 Long, E T
8 xxxi, Hewlett, R., The Gloucestershire Court of Sewers 1583-1642, Gloucestershire Record Series Vol 35, The Bristol and Gloucestershire Archaeological Society 2021
9 xxxii, Hewlett
10 xxxiv Hewlett
11 lxxvii Hewlett
12 3 Kissack, K., The River Severn, Terence Dalton Lit, Lavenham, 1982
13 21, Rhodes, J., The Severn Flood-Plain at Gloucester in the Medieval and Early Modern Periods, Presidential Address, trans BS & Glos Archeol vol 124 vol 124 ditto 2006
14 35 Presidential Address, trans BS & Glos Archeological Society, vol 124 Rhodes, ditto
15 162, Robinson, WJ., West Country Churches vol 2,
16 Wilshire, L, The Vale of Berkeley, Robert Hale, 1954
17 244, Adams, William, Adams Chronicle Of Bristol, J W Arrowsmith, Bristol, 1910
18 30 Hewlett
19 Abstracts and Extracts of Smyth's Lives of the Berkeley Family, manuscript
20 41 Kissack, K., The River Severn, Terence Dalton Ltd, Lavenham, 1982
21 56 Kissack
22 81 Kissack
23 268 Wilshire
24 15 Smith & Ralph
25 245 Finberg
26 84 Waters
27 94 Waters
28 81 Smith & Ralph
29 do 83
30 148 Waters, B., Severn Tide, Dent, JM, London, 1947
31 155 Waters, B.,
32 140 Kissack
33 84 Mac & Whitard, Mee, A., The King's England: Dorset, Hodder & Stoughton, London, 1967
34 14 Waters
35 36 203, Mee, A., ed Long, E T., ,The King's England, Gloucestershire, Hodder & Staughton, London, 1966
36 218 Mee, Gloucestershire
37 newnhamonsevern.co.uk
38 Victoria County History

**12 Severn Crossings**
1 15 Nicholls, T. F. & Taylor, J., Bristol Past and Present, Vol 2., J W Arrowsmith, Bristol, 1886&T Ecclesiastical
2 18 Waters, I., The Port of Chepstow, the Chepstow Society, Chepstow, Gwent, 1977
3 14 Kissack, K, The River Severn, Terence Dalton Ltd, Lavenham, 1982
4 405 Hutton, E., Highways and Byways in Gloucestershire, Macmillan and Co., London, 1932
5 22, Cooke, GA, Topographical and Statistical Description of the County of Gloucester, Sherwood, Neely, and Jones, London, undated

6 37 Robinson, W.J., West Country Churches, vol 1, Bristol Times & Mirror, Bristol 1914

7 Rhys, B., The South Wales Coast From Chepstow to Aberystwyth, County Cost Series, Fisher Unwin, London, 1911

8 155 Barber, C., ed Hando's Gwent, Blorenge Books, Abergavenny, 1987

9 30 Cooke

10 39 Finberg, H. P. R., The Gloucestershire Landscape, The Making of the English Landscape Series, Hodder & Stoughton, London, 1975

11 40 ditto

12 79 Waters, B., The Severn Tide, Dent, J., London, 1947

13 555-6, Seyer, Rev. S., Memoirs Historical and Topographical of Bristol and its Neighbourhood, From the Earliest Period Down to the Present time, Vol II, J.M. Gutch, Bristol, 1823

14 Wood, M.K., Smith, A.E., Newnham-on-Severn: A Retrospective, Gloucester 1962

15 443/4 Hutton, E.

16 Victorian County History: Lydney

17 556 Seyer

18 Sherborne, J.W., The Port of Bristol in the Middle Ages, Port of Bristol Series, Bristol Branch of the Historical Society, no. 13,The University, Bristol 1971

19 156 Barber, C., Ed., Hando's Gwent

20 157 Barber, C.

### 13 Gloucestershire Storms

1 21 Bigland, R, Original History of the City of Gloucester, John Nicholson, London, 1819

2 Parish Records, Almondsbury, Gloucester Record Office

3 14/5 Wilcox, Gloucester, 1590-1640

4 Tandy, D., And Did Those Feet, A Survey of Alkington for the Year 2001

5 do 310/11

6 268, Tandy

7 Wood, M.K.

### 14 Bristol and Its Hinterland

1 78, Napier, L., The Lower Severn Vale: Trade and Exploration in Tudor Times, Regional Historian

2 44/5, Latimer, J., Sixteenth-Century Bristol, J.W. Arrowsmith, Bristol, 1908

3 45 Latimer

4 260 Seyer, Rev. S.

5 ditto

6 lxxiii Hewlett, R., ed.

7 ditto

8 lxxiv Hewlett

9 p.19 E H Baker

### 15 Somerset Floods

1 15, Pool, S., So Much Loss and Misery: Taking the Long View of West Country Flooding, The Regional Historian, Issue 28 Spring 2014

2 158 Mee, A., Revised Long, E.t., Somerset: The King's England, Hodder & Stoughton, London, 1951

3 Camden, W., Britannia, or, A Chorographical Description of Great Britain and Ireland, together with the adjacent Islands

4 82 Robinson, W J, West Country Churches vol iv,
5 50 Robinson, W.J., West Country Churches vo 1,
6 39/40 Mee
7 53 Robinson, vol. 1
8 53 Mee
9 110 Pevsner, N., The Buildings of England, South and West Somerset, Penguin Books, London, 1958
10 Wikipedia
11 207/8 Mee
12 182 Mee
13 wiki
14 14 Gray
15 15/6 Gray

**16 Barnstaple and Beyond**

1 285 Mee, A., Ed., The King's England, Devon. Cradle of Our Seamen, Hodder & Stoughton, London, 1949
2 Hayton, D., Pamphlets on the Earthquake of 1580, dhayton.haverford.edu, July 17 2013
3 14 Norway, A H, Highways & Byways in Devon & Cornwall, Macmilan & Co., London, 1898
4 16/17 do
5 19 do
6 304 do
7 303/4 do
8 57 H&B D&C
9 14, Gray, T
10 15, do
11 56 Lois Lamplugh
12 240, Seyer

**Conclusion**

1 50 Davidson, Marshall B., Life in America, Houghton Mifflin Co., Boston, 1951, vol 1
2 39, Gethyn–Jones, E., George Thorpe and the Berkeley Company, Allan Sutton. Gloucester, 1982
3 42/3 ditto

# INDEX

**Aberavon**: beach 98
Abergwaitha: 51
Abergele: 3
Aberthaw: 7, 101; East 7
Adams, William: flood 1636 132
Adam the Fleming: 143
Adeliza of Louvain: 4
Adrian IV, Pope: 4
Alexandra Dock: 113
Almondsbury: 45; flood record, 158; Knole Park, 149
Alvington/Abone/Sea Mills, 152
Amsterdam, ship for, 98
Appledore: 187
Alvington/Abone: ancient oak trees, 142
Armada: v
Anglesey: ship lost off, 5
Apples: local varieties, 105
Arlingham: bizarre drowning, 151; burial delayed by 1607 flood, 136; church: 143; Church built flood walls, 130; Roman land reclamation 45; Sea Walls built, 137

Ashleworth: frequent floods, 152; flood marker in church, 152; ; stone from, 131

Aurelius, Marcus, 101

Aust: 32, 140; alabaster: 152; ferry/passage, 118, 140, 148, 150, 150; Bob Dylan, 148; Great Flood 1607, 59; New Grounds, 134;

Awre: ancient chest for drowned corpses, 142; land eroded, 136, narrowed above Awre, 153; whale sighted at, 137;

Aylburton: 142

**Barnstaple**: 41, 182; earthquake, May 1607; tempest, 1586

Barry: 101

Bassaleg/Bashalecke: 55

Bath: Earl of , 78

Beachley: 142; Head, 155; Peninsula 12;

Bell: Bristol made, 115

Benedictine priory: 113

Berkeley: Castle 45; Company, 194; Lord, built Arlingham sea walls, 137; Lord Thomas funded hermit chapel, 138; Pill, flood prevention: 160; Plantation: v; Somerset family: 47; Tower: destroyed: 159

Black Death: 14, 29

Blind man: 39, 87

Bores: Bridgwater, 126; Lougher Estuary, 126; Severn, 126; Taff, 126

Braunton Burrows: 187

Brean/Brian Down: 73, 175; Isle of Frogs: 176

Brent: flood victim 49; marsh 47

Bridges: between Gloucester and Bristol 32; Chepstow, 119

Bridgwater: 41, 80, 178; why so often flooded, 78

Bristol: 8, 33, 109, 164; Channel: 1, 47; Flood and wind,1484: 164; Great Flood 1607: 59; ships lost: 1484, 165; James, St, Fair: 21; Temple Fair 1754; greatest snowfall: 153;

Britannia: book, 174

Brunel: I.K.: Central Cardiff Railway Station, 103; Railway tunnel, 155

Bryant: Dr Ted, University of Wollongong: 92

Buckingham, Duke of, Water: 130

Burnham-on-Sea: 1607 flood, 73, church, 176

Burial: delayed by flood, 136

Burry Port: 95;
Burrows: Pembrey, 95; Towyn, 95
**Caldicot**: 55; Levels, Moors, 34
Camden, William: 174
Cardiff: Churches, St John, 103; St Mary's 38; Flood, 1968: 104; Norman settlers, 101; Viking raiders, 101
Caerleon: 4, 114, 140
Caerwent: 140; ford across the Severn: 149
Caldicot: church, 50; Levels 4, 44
Calais: 185
Calendar: Gregorian 27, 70; Julian 27, 70; local variations, 27
Caldicot Levels: 106
Caerwent: 149
Cannington: 177
Cardiff: Docks, 3; 1607 flood: 103-4; St Mary's Church: 60, 103
Carmarthen: 4; Bay 2, 94; -shire 34; 4 small children, 40
Cardigan: 34; -shire 34
Cary, River: 171
Castlemartin: 3
Cathedral of the Moors: Nash, 52
Cefn Mably: 49
Centurion Sartorious, 112
Chandos: Robert de 50, 111
Chapel Rock, 153; hermit, 153
Chepstow: 34, 118;
Christchurch: 54
Churches Conservation Trust: 94
Churchyard, Thomas: 183
Civil War, English: 25
Climate Change: 107
Clevedon: 173
cock boats: 59
Cold Knapp: marbles deposited in, 1607, 101
Cowbridge: 34
Curig, St.: 50
Cwm-Yr-Eglwys: 5

## 208 | Index

**Deerhurst** Priory: 113
Defoe, Daniel: 17, 24, 26, 110, 153; visited Aust, 153
Dinas, Pembrokeshire: 5
Dixton, floods: 121
Doggerland: remnants swept away: 70
Donne, John: sermon: 193; Virginia Company, 193
Drowning, bizarre: 151
Duke of Buckingham's Water: 129
Dunwich: 6
Durdham Down: Via Julia, 149
Dutch vessels: 98
Dylan, Bob, No Direction Home: 140, 148
**Earthquake,** Dover Straits: 183
Eels: 139
Elmore: land reclamation 45; meanderings stable, 130; sea wall repairs 1555
Emden, flooded 70
English Stones, 150
Epney, cider, 152; eel fishery, 139, 152; teazles, 152
Esterling, Le: 4
Eton College: 113
Evelyn, John: vi, 17
**Farmers**: 29
Fishing industry: 109, 136
Fitzhamon: Robert 4, 49
Fitzrobert, Mabel: 49, 52, 109
Flemish Hoy: 82
Flemings: 4, Adam; in Laugharne, 95; Philip le, 4; Joan le,101; Flemingston 4, 101; settled, 95
Flood: Great tide Berkeley: 1586; 160; Herefordshire 83; All Saints 70; of 1703: 118; St Marcellus: 6; Wye, 119; causing damage: 120
Floods, Bridgwater 1567, 78; Gloucester: record; in 1770, 124; Mylapor, 53; Noah, 1570, All Saints, 1636, 1707 , Storm, 1703; Shropshire, causing havoc: 1607, 1957; St John the Baptist 1258: 158 Wye 1735, 1793, 1852, 1947, The Great of 1606/7; total cost 48; counties: 29;
Flood Defences: Medieval: Frampton-on-Severn, 126; Saul, 126;

Slimbridge, 126; Fretherne, 126; Roman: Arlingham 127, Fretherne, 127, Hill, 127, Longney, 127;

Flying fish: 139

Forest, sunken: 3, lost off Cornwall: 185

Forest of Dean: Mines 149

Framilode: 135

Frampton-on-Severn, Court: 144; green, 143; land reclamation, 45; passage, 150; tower: 159; water level, 134

Fremington, Vicar of: records of extreme weather: 188

Freshwater Bay: 2

Frisia, flood: 70

Fretherne, land reclamation 45

Frost, Bristol November 1607 61/2

Fust, family: 23, Jenner Fusts 24; sea grass 136

**Gdansk**: grain imported from: 188

Gentleman's Magazine, 60

Gentleman of Worth: 36

Gerald of Wales: 2, 5, 152

Germans: eel consumption, 139

Glasbury village: 120

Glastonbury Abbey: drainage works, 171; Lake Villages 1; Tor 93

Gloucester: lowest river crossing, 149

Goldcliff: 34, 50, 112, 113; brass plaque memorial 88; priory: 15, 107; sea wall built, 112

Gout: The Great, 109

Goodwin Sands: 17

Gower Peninsula: 2, 4, 5

Grantham: over-flown 73

Gray, Bishop: 149

Great Drowning of Men/Grote Mandrenke: 6

Great Flood: 69; losses from, 133; Norfolk 69: Somerset 69

Great Gout: 109

Great Storm of 1703: 24

Great Winter: 1683/4, 153

Green Grounds: 5, 97; inner, 5; Outer, 6

Green Man: corbel, 115

Gregorian dating system: 18
Grounds: Cardiff, 108
Gunpowder Plot: 28
Gwent Wildlife Trust: 51
**Hailes** Abbey: 2
Harleian Library: 87
Harvest, 1607; failures 1587, 1588, 1589
Hasfield: regular floods, 152
Haslett, Prof Simon: 92
Henbury: ancient settlement, 131; 1607 flood, 59; parish, 140; salt marshes/Salso Marisco drained 45;
Hereford, floods
Herpath: 177;
Henry VIII, King: 18
Herbert family: 47; Lord, 88; of Penarth, 102
Hereford:, floods: 1795,1947, 1960: 83, 121
Hermit: 142, 153; hermitage, 2
Hill: Great Sewer at pill, 23; St Michael's parish, 23; flats/waste, 138; sea wall built, 45; wall repairs, 129;
Holm: Flat, 74; Steep, 74; Great Flood 73;
Hoy, Flemish: 82
Hundred Years' War: 137
Huntspill/Huntsfielde: 176; flooded: 73
**Ice Age**: Mini vi
Ice Fairs: vi
Irish Rebellion: 1798, 98
Ironbridge: 124
**Jews**: Bridge tolls 135
Joseph of Arimathea: planted his staff, 138
Julian, dating system: 18
**Kenfig**/k: 6; Burrows 2; Pool 6, 7, 100; town: 98
Kidwelly: 95
Ken More: 173
King's Sedgemoor Drain: 171
Kingston Seymour: 67, 173
kypes, 138

**Lambeth**: 49, 51
Lancaut Peninsula: 118, 142
Lanckstone: 50
Lands: swept away: Arlingham, 127; Fretherne, 127; Hill, 127; Longney, 127
Laugharne: 5, 34, 95
Leland, John: 14, 99
Levels: Gwent 14; Somerset
Lewis family of St Pierre: 118
Liberation of Zeeland and Holland, 70
Lisbon earthquake: 92
Littleton-upon-Severn: 141, 145; flood banks maintained, 139
Llandaff: Bishop of, 51; Mistress Matthews of, 88
Llanfihangel: 50
Llantwit Major, 100
Llanwern: 14, 51, 93
Longitude Prize: 17
Longney: land reclamation 45
Lost Land of Lyonesse: 185
Louvain, Adeliza of, 4
Lovelock-Evans, Graham, 'The Romans in Barry", 101
Lyme Regius: landslips 184
Lydney: salt marshes, 153; reclamation of salt marshes, 142
Lysaght Memorial, 116

**Magor**: 51, 115; green man corbel, 115; salt marshes reclaimed 51; Pill: 51
Maisemore: bridge over the Severn, 149
Malaga: Battle of ,17
Malago River: Somerset, 60
Malmesbury, William of: 161
Margam Abbey: 2, 100
Marshfield: 49, 110
Matherne: 34, 51
Mendip Lead, 154
Merthyr Mawr, Nature Reserves: 2, 7, 98
Midsummer Night's Dream, A: v

Minehead: 178; trade with Virginia & West Indies: 179

Mini Ice Age: vi, 6

Minsterworth: church built flood walls 130; flood: 'thorough tempest', 160; floor raised by 4 feet: 130;

Monmouth: 121

Montague, Sir Walter: 88

Moors: Penarth, 108

Morgan family: 111

Mortehoe: 183

Morris, Sir John 97

Montford: bridge tolls 135

Mumbles Head: 5,6; Hill, quarry on 97

Mumby Chapel: 70

Mylapore: 53

**Napoleonic** Wars: Chepstow, 118

Nash: St Mary church: 16, 51; print of flood, 48

National Library of Wales: document, 27

Netherlands: 6

Newgale: 2

New Grounds/Warth: 134, 144, 153, 194

New Passage/Crossing: 45, 150, 153; flood risk, 127; Binn Wall, 45

Newnham-on-Severn: crossing at low water, 151; erosion, 151; moved to higher ground, 136, 151; violent floods, 151, 161

Newport: 34; Wetlands Centre, 16

Newton: 109; 1795 flood; Nottage 6, 98

Nodens: Temple of Celtic god: 149

Norfolk: 66

Normans: 14

Northleigh: rector poet, 183; Dover Straits Earthquake of 1586: 183

Northwick: 45; flood, 140

Norton, John, architect/engineer: 51

Norway: flood

**Old oaks** visible: Alvington, 127

Oldbury-on-Severn: 145: dispute over sewers, 45; Roman camp: 145; Salmon pool, 137; The Toot, 145

Oldbury Mills: blamed for flood, 133

Old Passage: 149
Otterhampton: 178
Over: Roman crossing, 149; Telford's bridge, 149; rebuilt, 131
Oxwich, 96; Bay: 5 Parsonage House
**Paignton**: 187
Parrett, River: 125, 170, 178
Parish Records: 19
Pawle, John vicar of Almondsbury, 158
Pawlett: 176; hams, 176
peat beds: Port Talbot, 3
Pembrey Burrows: 95
Penarth: 48, 102; Moors, 37
Penmaen: Church, 98
Pennard: 2, 5
Pilney: 45
Pilning: 141
Pirate: William Chick, 148
Plague: 20; Bristol 1603; of Pestilence 28;
Plynlimon, river: 34; rainfall, 130
Plowmen: 29
Porpoises: 138
Port Talbot: 2, 3
Portbury: 154, 172
Porthcawl, drowned sailors buried, 98
Porthkerry: 7
Portishead: 173
Portskewett: 22, 51, 116; Roman port, 116; Severn ferry terminus, 116
Prayer, fisherman: 179
Poured landscape, 125
Purton, ferry, 144
**Redwick**: Monmouthshire: cider and orchards 115; Early, apple variety, 115; ferry: 153; Open Air Museum, 114; St Thomas the Apostle church, 114; South Gloucestershire: 133; 141
reens: 107
Road, for ships: 108; whale 106

214 | Index

Rockingham: flood record: 134
Rogiet: 53
Roman: artefacts, 4, 51; in Cardiff, 101; embankments, 107; fords, Caersws, Forden 130; Invasion, port
Rhymney/Rumney: 14
Romney: Flood victim, 49
Romans: Invasion 7; Port 14, 22
Rotterdam:13
Rumney/Rhymney: Great Wharf, 18, 109; River 114
Rupert, Prince: 150
**St Augustine's** Abbey Bristol: 21, 23, 48
St Bridget/Bride: 's Bay: 2; Brean Down, 74, 175; Netherwent 54, 111; Wentloog, 14, 16, 110
St Curig: chapel 50
St Ishmaels: 96
St James' Fair Bristol: 21
St John the Baptist Bristol: 20
St Marcellus Flood: 6
St Mary-le-Port Church, Bristol: 20
St Mary Magdalene, Goldcliffe: 16
St Mary Swansea: 27, 110; poorly documented 98
St Mary the Virgin: Nash 16 ; Portskewett, 52
St Mellons: 14, 54
St Nicholas church, Bristol: 19
St Peter: Peterstone 14, 18, 21,52
Sts Phillip & Jacob church, Bristol: 20
St Pierre: 22, 54, pill, 150
St Tecla's/Thekla Island/chapel: 118, 142
St Thomas the Apostle, Redwick: 15, 34,53 Monmouthshire; bells 53 South Gloucestershire: 45
St Twrog's chapel: 118
Salmon pool: 137
Salt Marshes: 51; Henbury, 128, 154; Roman 51
Sandhurst: frequent floods, 152
Sand Bay: 174
Saul: land reclamation 45; 144

Sea horses: 139
Seaton: 184
Sea Mills: 153
Seawall Tea Rooms: 15
Severn: bargees, drunk 151; Beach: 141; 45 Bores; Crossings 4, 12; Floods: 1636, isolated by, 132; various, 161; Frozen,153; River vi, 1, 7
Sewers, Court of: 45, 133
Seyer, Reverend S.: 19
Shakespeare, William, The Tempest: 193
sharks: 138
Shepherds: 29
Sharpness Canal: 144
Ships, road for: 5
shipwreck memorial of 1735, 139;
Shirehampton: 19
Shrimp, 138
Skenfrith: 16
Slimbridge: flood banks maintained, 139; land reclamation 45, 136; New Grounds, 144; Wetlands Centre, 144
Somers, Admiral Sir George: 193
Somerset Levels: 93
Spanish Armada: 8; Succession, War Of , 17
Spring Tide floods: Severn, 1687/8, 1688/9, 1703, 128
Statorius, plaque by: 114
Steart Marshes Wetlands Reserve: 177
Stockland Bristol: 177
Storm: The 110, book 17, surge, of 1584, 101
Sudbrook/Sudbury: Camp 22, 150, 155; Captain Blethin Smith burial 22, 155; pill, 150; sea wall, 114; Trinity Chapel ruin 22, 155
Sully, smallest castle in Glamorgan 101; off Cornwall, 186
Sunken Lands: 3
Surinam, ship from, 98
Swansea: 34; Corporation records: 89
**Teignmouth**: 185
Telford, Thomas: extreme weather 131; highest flood, 131; Low Water, 1796 135

Temple Fair Bristol, 80
Tewkesbury Abbey: floods 1484, 1587, 1611: 159; flood maker, 130;
Thames: 125
Thorpe: George, of Berkeley: 194
Tidenham: 142
Tintern: iron works, 120
Tiverton: houses destroyed by floods; 189; flood warnings by schoolboys: 186/7
Tirley: floods, 139; flood marker 1924, 139
Towey Estuary: 94, 95
Tredegar House: 52, 111
Trinity House, records: landmark, 138
Tsunami: 4, 92
**Ubley**: 21
Undy: 51, 115
Uphill: 175
Upper Arley, ironbridge: completed in record time, 135
Usk: 14
**Via Julia**: 2
Victorian County History: 15, 19
Viking raiders, Cardiff, 101
**Warth**: 14
Watchet: 178
Welford: v,
Wentloog/Wentwood: 4, 13, 14 49: largest forest in England, 44; Levels 21, 44
Westbury-on-Trym: 19
Weston-super-Mare: 174; Birkbeck Island, 141; Gazette Office, Gull yellers, 141; museum 32
Weston Zoyland: 178
whale: seen at Awre, 137
Whitchurch: drowned brothers 1853, 122
Wh(i)tson: church 15, 113
Wilfrck: 54
William Chick, pirate
William of Worcestre: 148

Wiston: 5

Witney-on-Wye: 120

Wizo: 5

Wood, Richard: weather records 188

Woolaston Chapel, landmark: 138

Worcester Floods, 1607, 161; flood levels on walls: 130; William of, 148

Wye: River, 34, 154

Wyatt, Adam: 187

Yatton: St Mary's, 173

**Zeider** Zee: 6

# ABOUT THE AUTHOR

Barb Drummond has been researching and publishing books on British history for several decades, specialising in 18th century history.
    She can contacted on @barbdrummond.bsky.social

        facebook.com/Barb%20Drummond%20historian
        x.com/Barb_Drummond
        amazon.com/BarbDrummond

Milton Keynes UK
Ingram Content Group UK Ltd.
UKHW051058280824
447506UK00010B/19

9 781912 829170